Springer Theses

Recognizing Outstanding Ph.D. Research

Aims and Scope

The series "Springer Theses" brings together a selection of the very best Ph.D. theses from around the world and across the physical sciences. Nominated and endorsed by two recognized specialists, each published volume has been selected for its scientific excellence and the high impact of its contents for the pertinent field of research. For greater accessibility to non-specialists, the published versions include an extended introduction, as well as a foreword by the student's supervisor explaining the special relevance of the work for the field. As a whole, the series will provide a valuable resource both for newcomers to the research fields described, and for other scientists seeking detailed background information on special questions. Finally, it provides an accredited documentation of the valuable contributions made by today's younger generation of scientists.

Theses are accepted into the series by invited nomination only and must fulfill all of the following criteria

- They must be written in good English.
- The topic should fall within the confines of Chemistry, Physics, Earth Sciences and related interdisciplinary fields such as Materials, Nanoscience, Chemical Engineering, Complex Systems and Biophysics.
- The work reported in the thesis must represent a significant scientific advance.
- If the thesis includes previously published material, permission to reproduce this must be gained from the respective copyright holder.
- They must have been examined and passed during the 12 months prior to nomination.
- Each thesis should include a foreword by the supervisor outlining the significance of its content.
- The theses should have a clearly defined structure including an introduction accessible to scientists not expert in that particular field.

Roberto J. Galván-Madrid

On the Formation of the Most Massive Stars in the Galaxy

Doctoral Thesis accepted by Universidad Nacional
Autónoma de México

 Springer

Author
Dr. Roberto Galván-Madrid
National Autonomous Univeristy
    of Mexico (UNAM)
Harvard-Smithsonian Center
    for Astrophysics (CfA)

Supervisor
Prof. Luis F. Rodríguez
Centro de Radioastronomía y Astrofísica,
UNAM
México

ISSN 2190-5053   ISSN 2190-5061 (electronic)
ISBN 978-1-4614-3307-1   ISBN 978-1-4614-3308-8 (eBook)
DOI 10.1007/978-1-4614-3308-8
Springer New York Heidelberg Dordrecht London

Library of Congress Control Number: 2012938195

Printed on acid-free paper

Springer is part of Springer Science+Business Media (www.springer.com)

This document has been elaborated by Roberto Galván-Madrid and presents a compilation of the research performed to obtain the degree of Doctor in Science (Astronomy) from the National Autonomous University of Mexico (UNAM).

Muenchen, Germany   Roberto Galván-Madrid

Preface

Star formation, the process that transforms diffuse gas into stars, is one of the most active research topics at present. We are concerned with understanding the role of star formation both as a function of time as well as across the range of stellar masses. The monotony of the initial Universe was broken by the formation of the first stars, only a few hundreds of millions of years after the Big Bang. These first stars reionized the surrounding gas and started cosmic complexity.

In spiral galaxies like ours, where interstellar gas is still relatively abundant, the formation of new stars is an on-going process that we can study and try to understand better. Large advances have taken place in the research of the formation of solar-type stars, those similar to our Sun, possibly because they are so abundant and relatively easy to observe. However, the formation of low-mass stars and brown dwarfs on one side of the stellar mass spectrum and of high-mass stars on the other, remains poorly understood.

It is on the formation of the most massive stars in the Galaxy that this work is centered. Different regions of massive star formation are studied in detail, using the powerful techniques offered by radio interferometry. Massive stars are formed and spend all of their youth inside dense regions of gas and dust and the classic methods of optical astronomy cannot be used. The different processes that are studied in the radio are described in a brief manner and the results presented in a clear, straightforward way.

Interestingly, the results of Galván-Madrid suggest that the formation of the most massive stars is consistent with a scaled-up version of the formation of solar-type stars. In both cases, there is gravitational infall of gas from a core into an accretion disk, where the gas slowly loses angular momentum to be able to finally reach the star. Nature's way to permit accretion involves the production of powerful outflows that remove excess angular momentum from the gas in the disk. As in solar-type stars, disks and outflows appear to be the key ingredients in the formation of massive stars. Another interesing result from this thesis is the evidence found for a larger influence of the environment outside the star-forming core than previously recognized.

The evidence, however, is far from conclusive. Massive stars are remote, heavily obscured and form in clusters; all of this complicates the observations. Fortunately, there is a new generation of interferometers that will allow a much more advanced understanding of massive star formation and the possible confirmation that they form like solar-type stars, but in a big way. Of course, the possibility that something else is lurking in that darkness is still open.

Centro de Radioastronomía y Astrofísica (CRyA)   Luis F. Rodríguez
Universidad Nacional Autónoma de México (UNAM)

Acknowledgements

A lot of people have been important for the completion of the work presented in this thesis. I am deeply grateful to my advisor at CRyA-UNAM and mentor Luis F. Rodríguez. I worked with him for the first time as a summer undergraduate in 2003. Since then, he keeps teaching me valuable things for work and providing personal encouragement.

During most of the thesis work, I was based at the Harvard-Smithsonian Center for Astrophysics (CfA) as an Smithsonian Predoctoral Fellow of the Submillimeter Array (SMA) project. The CfA has a challenging intellectual environment and it has been a pleasure to work in such a place. I have been greatly enriched in talks, lunches, and conversations with fellow students, postdocs, and senior staff. I am indebted to my CfA advisors Qizhou Zhang and Paul Ho. Qizhou has always been there to answer even the smallest question, and Paul has provided me with encouragement and interesting research ideas. I am also grateful to Eric Keto, who has been a very important collaborator since I arrived to the CfA. I also want to thank Nimesh Patel, Mark Gurwell, Ray Blundell, Ken "Taco" Young, Shelbi Hostler, Erin Brassfield, Anil Dosaj, Ryan Howie, Jennifer Barnett, Margaret Simonini, Steve Longmore, Ke Wang, Thomas Peters, Mordecai Mac Low, Karin Hollenberg, Stan Kurtz, and Enrique Vázquez.

I could not have done this thesis without the distant but continuous support from my friends from life. Thank you, Karina Arjona, Nila Chargoy, Gabriela Montes, Rosy Torres, Martín Ávalos, Jesús Toalá, Vicente Hernández, Carlos Carrasco, Karla Álamo, David Medellín, Raúl Lamadrid, Eduardo Montemayor, and Arturo Montemayor. I am also grateful to the friends I have made in Cambridge/Boston, Jaime Pineda, Diego Muñoz, Arielle Moullet, Vivian U, Sophia Dai, Jan Forbrich, Sergio Martín, Alexa Hart, Dave Riebel, Mayumi Sato, Natasa Tsitali, Katharina Immer, Molly Wasser, Apple Hsu, Ibon Santiago, and last but not least, Baobab Liu. Most of this thesis was assembled while visiting Academia Sinica in Taiwan. Thanks to Satoko Takahashi, Anli Tsai, Pei-Ying Hsieh, Huan-Ting Peng, and Alfonso Trejo for a wonderful stay.

Finally, I want to thank my mother, I wish I could be at home more often, my brothers Pedro and Andrés, and the rest of my family. Thanks to my uncle Pepe for supporting me during my college years when I started this adventure in Physics, I hope you get better soon. To my aunt Eloisa, to Brenda, and Valeria.

Contents

Chapter 1
Introduction to the Scientific Problem

1.1 Star Formation

1.1.1 Stars in Context

Stars are the basic building blocks that drive the evolution of the Universe. Although dark matter and dark energy possess the largest share of energy-matter content at present – 23% and 72% respectively [192] – the former only interacts gravitationally and its effects are negligible on scales relevant to the interior of galaxies, while very little is known about the latter. On the other hand, the much smaller share of "normal" atoms in the Universe is the main agent responsible for driving its evolution.

Star formation, the process by which gas is transformed into stars, is one of the two reciprocal agents that control the evolution of galaxies [144]. The other process is feedback from the formed/forming stars: gas is transformed into stars that through outflows, winds, H II regions, and radiation shape the galactic ecosystems that will form the next generation of stars (Fig. 1.1). The more massive the star is, the greater the feedback is. The first generation of stars that formed from primordial elements (H, He, and Li) in the Universe was responsible for its reionization during redshift $z \sim 5$ to 10 [1, 34]. Heavier elements are formed either by nucleosynthesis in the interior of stars or from supernova explosions [37]. Elements heavier than oxygen up to iron are synthesized exclusively by the most massive stars, while elements heavier than iron are synthesized in supernova explosions (Type II and Type Ib) [215], which are the explosive ending of the rapid life of a massive star. Therefore, the composition of the Universe as we see it is a byproduct of stars, and in particular the heavy elements so necessary to life could not exist if it were not for massive stars.

The process by which stars form is itself deeply connected to life. During their formation, stars gain mass through accretion from a disk of dust and gas that results

R.J. Galván-Madrid, *On the Formation of the Most Massive Stars in the Galaxy*, Springer Theses, DOI 10.1007/978-1-4614-3308-8_1,
© Springer Science+Business Media New York 2012

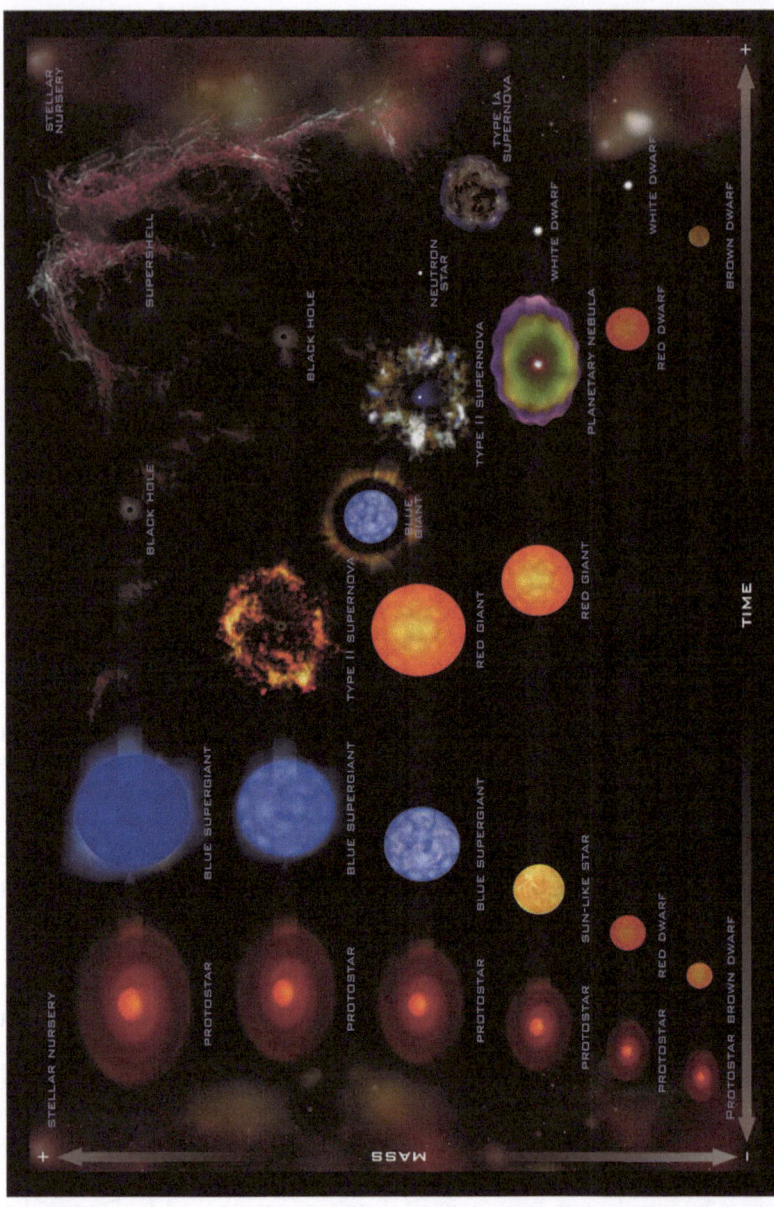

Fig. 1.1 Illustration of stellar life cycle as a function of time (x-axis) and mass (y-axis). Stars of all masses are formed from the collapse of interstellar gas. During their lives and deaths they synthesize elements that end up forming part of new gas clouds that form new generations of stars. Credit NASA/CXC/SAO

from the collapse of the parental "core" [144, 184]. It is in these dusty disks that planet formation occurs. Therefore, star and planet formation are intimately tied to one another.

1.1.2 From Clouds to Clumps to Cores

Stars form within Giant Molecular Clouds (GMCs) that have sizes up to several $\times 10\,\mathrm{pc}$[1] and masses up to several $\times 10^4 M_\odot$. GMCs have multiple levels of substructure and it has been found that at the sites relevant to star formation, smaller scales correspond to denser, more quiescent structures, at least before important feedback has taken place. Table 1.1 summarizes the basic terminology and scales relevant to our discussion.

Larson [130] found the following scaling relations between density ρ, size R, and velocity dispersion δv in GMCs:

$$\rho \propto R^{-1}, \tag{1.1}$$

$$\delta v \propto R^{1/2}. \tag{1.2}$$

The second relation (1.2) has been more widely confirmed than the first one, and it is naturally expected if GMCs are a turbulent, compressible fluid subject to shocks [144]. The role of turbulence in star formation is believed to be twofold [16, 138]: on the one hand, it supports GMCs against global collapse; on the other hand, it is responsible for creating localized density enhancements that constitute the clumps and cores from which stars form [72, 207].

Once a local density enhancement becomes gravitationally bound it collapses into a point in a "free-fall time" t_{ff} in the absence of any other opposing force:

$$t_{\mathrm{ff}} = \left(\frac{3\pi}{32G\rho} \right)^{1/2}, \tag{1.3}$$

However, more factors are usually at play, in particular thermal/turbulent support at core/clump scales and magnetic fields. For the formation of low- and intermediate-mass stars ($M_\star < 8 M_\odot$, see Sect. 1.2) it appears that at small-enough scales ($R < 0.1\,\mathrm{pc}$) cores are reasonably well decoupled from their environment, and that magnetic [80] and thermal pressure [128] are important in delaying collapse. The turbulent linewidths at core scales for low-mass star formation are subsonic, which is often interpreted as turbulence being unimportant at core and smaller scales. However, there is still considerable debate about the relative importance of these

---

[1] 1 parsec (pc) is equivalent to 206 264.806 astronomical units (AU), or to 3.086×10^{18} cm.

Table 1.1 Properties of star-forming clouds, clumps, and cores[a]

Property	Clouds[b]	Clumps[b]	Cores[b]
Mass ($M_\odot$)	$10^3–10^4$	50–500	0.5–5
Size (pc)	2–15	0.3–3	0.03–0.2
Mean density (cm^{-3})	50–500	$10^3–10^4$	$10^4–10^5$
Velocity extent km s^{-1}	2–5	0.3–3	0.1–0.3

[a] Adapted from Bergin and Tafalla [25]
[b] Values are typical for low-mass star formation regions before significant feedback has occurred. They scale up for high-mass protostars (except for core and clump size)

agents in collapse [16]. One of the results of our thesis is that for the case of the most massive stars, star-forming cores do not appear to be fixed during protostellar accretion, but may continuously accrete from their environment.

More recently, another ingredient has been recognized as important in the picture for the initial conditions of star formation: the ubiquity of filaments (see Fig. 1.2). The advent of Galactic-scale surveys at infrared (IR) and millimeter (mm) wavelengths has shown that most often, star-forming cores are embedded in structures that are filamentary [148]. Moreover, sensitive observations of low-mass protostars have shown that also at core scales there are significant departures from simple spherical or axisymmetric structures [201]. All this evidence points toward a new picture of star formation in which the initial conditions are non-equilibrium structures.

1.1.3   Disk-Jet Mediated Accretion

On scales smaller than 0.01 pc and once a protostellar object has formed, a wide body of observations and theory has convincingly shown that low-mass stars gain mass through a circumstellar accretion disk [173, 185]. During the stages in which accretion is strong, a jet is often observed perpendicular to the disk [172]. Such a jet is launched via magnetohydrodynamical processes [38] and entrains part of the circumstellar envelope producing spectacular molecular outflows [9]. There is increasing evidence that this picture partially applies to the formation of massive stars (see Fig. 1.3). One of the results of this thesis is that for the most massive stars the presence of significant ionization must be taken into account.

During the most embedded stages of the formation of low-mass stars (class 0 and Class I protostars, see Adams et al. [3] and Ward-Thompson et al. [213]) it is difficult to disentangle the emission from the envelope and from the disk, but the advent of (sub)millimeter interferometers has alleviated this situation [101]. The least embedded Class II protostars and T Tauri stars exhibit naked disks that can be characterized in exquisite detail via molecular-line spectroscopy and resolved imaging in the mm [6]. Another valuable tool to characterize disks in low-mass

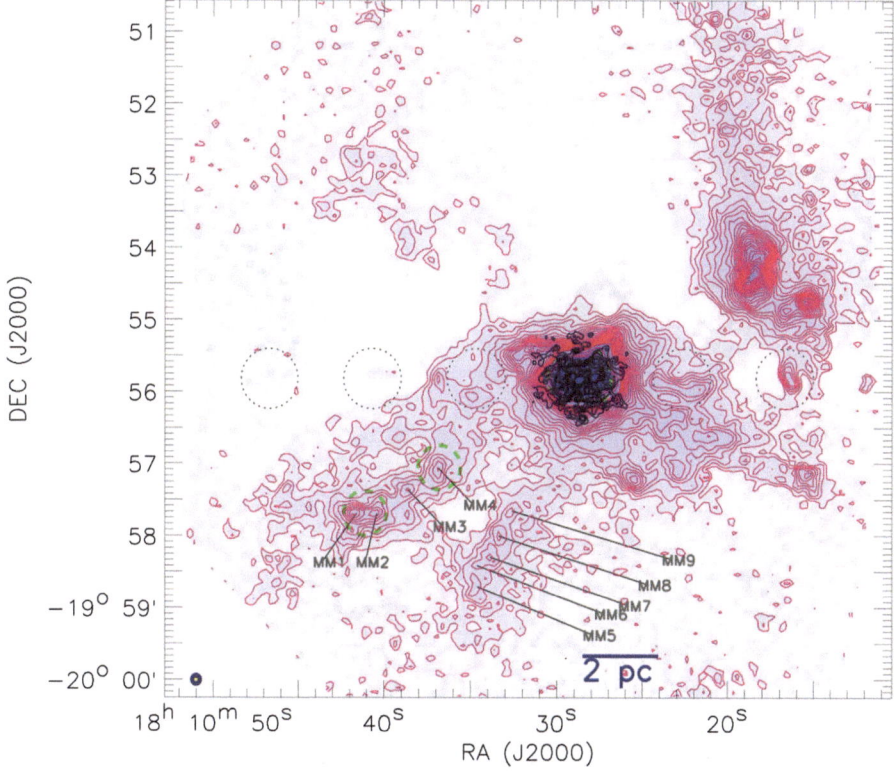

Fig. 1.2 Filamentary structures at scales from ~1 to ~10 pc in the massive star formation region G10.6 − 0.4. *Magenta contours* show the dust continuum at 1.2 mm (IRAM 30 m telescope). *Blue contours* show the combined single dish (IRAM 30 m) and interferometer (SMA) high-resolution image of the central cluster. Several massive clumps in the peripheral filaments are labeled. Credit Liu et al. [132] (Reproduced by permission of the AAS)

protostars is IR spectroscopy. While mm observations trace the bulk of dust and gas, the near (NIR) to mid (MIR) infrared are the main tools to characterize the warm dust in the inner accretion disk [53].

Accretion onto the stellar surface is believed to be mediated by the stellar magnetosphere [185]. The stellar magnetic field truncates the inner disk and accretion from the disk onto the star proceeds through the magnetic-field lines. The accretion shock formed at the stellar surface dissipates copious amounts of energy in X-rays and the UV. Recombination lines such as Brγ are widely used to estimate accretion rates for low-mass protostars [152]. Unfortunately, these lines are not detectable for deeply-embedded, massive protostars, so the way in which material is finally accreted by a massive star remains an unexplored topic.

More recently, IR surveys of large numbers of protostars have revealed that low-mass protostars are less luminous than they are expected to be from models and

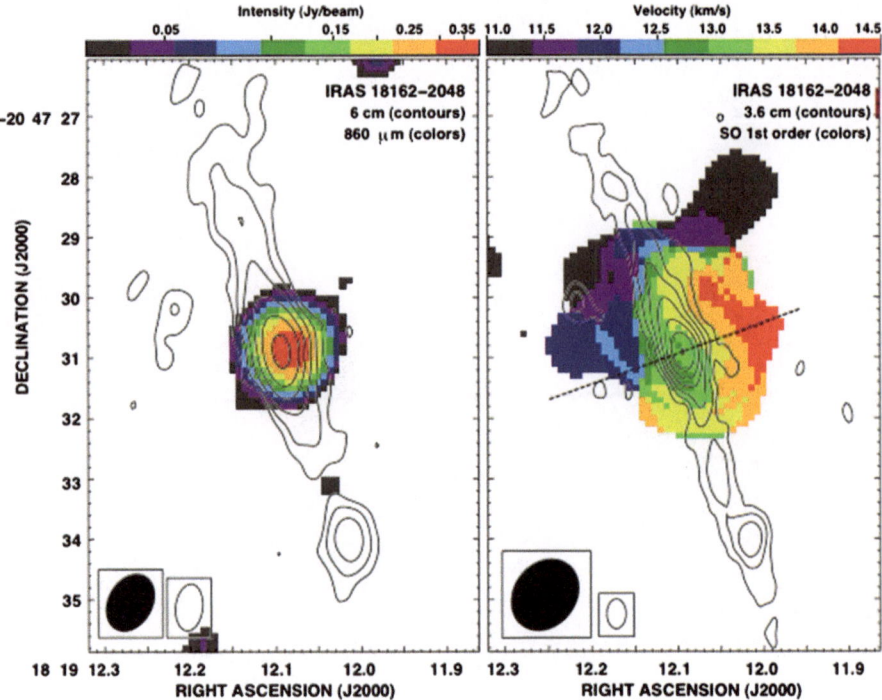

Fig. 1.3 Disk-jet system in the high-mass protostar IRAS 18162–2048 (HH 80–81). The *left panel* shows the envelope+disk traced by dust continuum at 860 μm (*color*) and the free-free radio jet seen at 3.6 cm (*contours*). The *right panel* shows in color the velocity field of the rotating disk+envelope traced by the SO molecule overplotted on the radio jet. Credit Carrasco-González et al. [40] (Reproduced by permission of the AAS)

our knowledge of their timescales [62]. The most appealing solution to this "low-luminosity problem" is episodic accretion [61], i.e., that protostars spend a large fraction of their lifetimes in a low-accretion mode with interspersed large accretion bursts. Evidence for variable accretion in a selected number of sources also comes from multi-epoch observations of free-free radio jets, both for low- and high-mass stars. These jets show variability of the jet cores and motions of the knotty ejections [48, 71, 142]. Since outflow and accretion are thought to be correlated, variable accretion seems to be a key element in the formation of stars of all masses. Detailed numerical simulations and models also show that when non-axisymmetric effects in 3D and disk instabilities are taken into account, variable accretion appears to be universal [122, 161, 237]. One of the results of this thesis is that variability in the inner ionized gas (the so-called hypercompact H II region) surrounding protostars with masses larger than about 15–$20 M_\odot$ appears to be a signature of active and variable accretion.

1.2 Massive-Star Formation

1.2.1 Properties of Massive Star Formation Regions

Few massive stars are born compared to their low-mass counterparts. The number of stars $N(M)$ that are formed within a mass range dM is given by the initial mass function (IMF) of stars [179]:

$$N(M)dM \propto M^{-\alpha}, \tag{1.4}$$

where $\alpha = 2.3$ for stars more massive than about half a solar mass. Per unit of mass range, for every $30 M_\odot$ star that forms, 13 stars with an initial mass of $10 M_\odot$ and 3,000 solar-mass stars are born.

Typically, studies aimed at the formation of massive stars need to target regions that are located at distances of several kpc from the Sun. Two other factors that conspire to hamper the study of massive star formation (MSF) are that massive stars form within more clustered environments than low-mass stars [127, 238], and that massive protostars are more deeply embedded, making the use of radio, (sub)mm, and IR techniques necessary for their study.

Several theories have been put forward to explain the properties of massive star formation regions (MSFRs). The IMF is probably the key aspect they all try to explain. One such theory labeled as "core accretion" or "monolithic collapse" assumes the existence of a massive prestellar core that will end up forming a single massive star [121, 145]. This model was mainly motivated by observations of the distribution of masses of low-mass prestellar cores that indicate it follows a power law with a slope close to that of the IMF (1.4) [150]. A second popular model for MSF and the origin of the IMF is so-called "competitive accretion" [18, 31]. In this model the distribution of core masses is not so important because cores are not well decoupled from their environment. Rather, the molecular cloud collapses in a lot of fragments with a Jeans mass:

$$M_J \propto T^{3/2}/\rho^{1/2}, \tag{1.5}$$

where T and ρ are the gas temperature and density respectively. Once this initial fragmentation is set, the accreting protostars compete with each other for their common gas reservoir.

Much research by numerous groups has been devoted in the last decade to test the above mentioned models (see Zinnecker and Yorke [238], and references therein), and models have been refined as well (see McKee and Ostriker [144], and references therein). Although still not fully recognized, models are converging to a compromise point. One key improvement has been the inclusion of radiative and gas feedback and the treatment of radiative transfer in numerical simulations [50, 121, 122, 161, 212], which yields a more realistic cloud fragmentation at different M_J for different subregions, unlike that expected from simpler versions

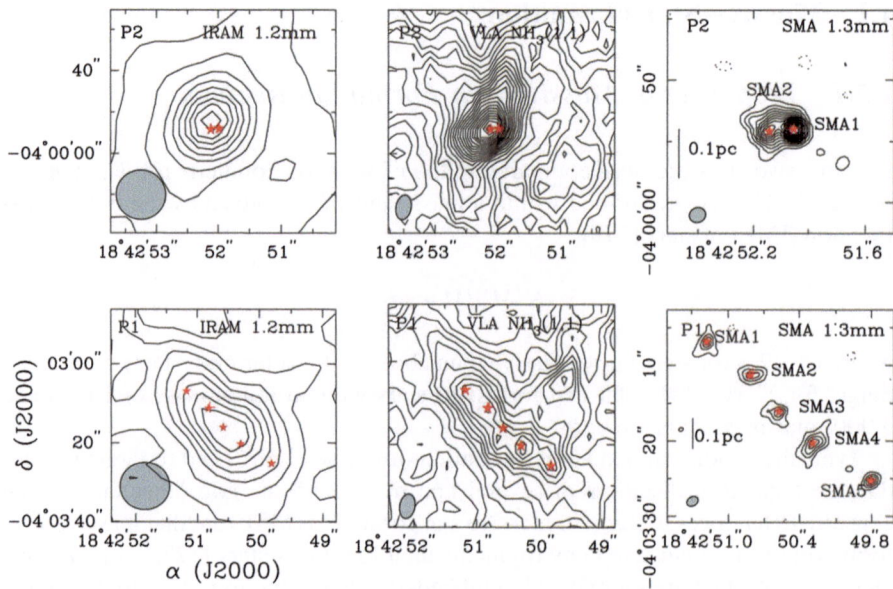

Fig. 1.4 Fragmentation in the massive star forming cores within the infrared dark cloud (IRDC) G28.34 + 0.06. The dust emission that in single-dish observations appears as a single core is resolved into multiple components. Credit Zhang et al. [236] (Reproduced by permission of the AAS)

of competitive accretion. Similarly, the more recent core-accretion simulations show that feedback limits, but does not suppress, fragmentation at core (see Table 1.1) and disk ($\sim$1,000 AU or less) scales [122], unlike the original motivation for a single-core to single-star scenario.

Observations, some of them presented in this thesis, impose important constraints on these models. Some cores detected in single-dish surveys have been thought to be progenitors of single massive stars because they have masses of a few $\times 100 M_{\odot}$, however, when observed with interferometers (Fig. 1.4) they break into several cores with masses of at most several $\times 10 M_{\odot}$ [75, 231, 236]. Also, observations show that at least for some very massive SFRs, the cores are still embedded in larger-scale accretion flows (clump and cloud scale) that appear to be undergoing global collapse and keep feeding the cores they harbor [74, 103, 133]. Therefore, the cores that form massive stars cannot be considered isolated.

1.2.2 Disk-Outflow Accretion in Massive Star Formation

It is well established that in low-mass protostars accretion proceeds through a circumstellar disk [6, 173] accompanied by the ejection of magnetically driven jets

[38, 172] that entrain the material seen as outflows [9] (see Sect. 1.1.3). In the past decade, evidence has gathered to assert that similar processes are present in the formation of high-mass stars, but at the same time there seem to be some important differences [30, 45].

It appears that outflows from massive protostars are wider and less collimated, although this may be partly due to the poorer angular resolution of observations compared to nearby low-mass SFRs [9, 26]. If massive outflows are not collimated, it is possible that they are not magnetohydrodynamically launched as their low-mass counterparts, however, a few observations have found that magnetic fields are important in massive star forming cores [81] and a magnetized jet has been directly observed in one MSFR [38]. It is still unknown if this is widespread in MSF. Also, it has been found that massive outflows increase their energetics (mechanical luminosity L_{mech}, mechanical force F_{mech}, and mass-loss rate $\dot{M}$) with increasing luminosity of the powering source(s), L_{bol}, following the same power laws as low-mass outflows [28, 134, 223]. This suggests that the launching mechanism is the same for protostars of all masses.

The free-free emission from radio jets at the base of molecular outflows has also been observed for some high-mass protostars with up to $M_\star \sim 15$ to $20\,M_\odot$ (corresponding to B-type stars) [48, 175] (see Fig. 1.5). When detected, this emission appears to trace collimated jets as in low-mass protostars. However, it is not clear how frequent these radio jets are in MSF regions, and what is their relation with other sources of ionization like the UV-photon induced ionization expected for masses larger than about $15\,M_\odot$ (see Sect. 1.3).

The case for the existence of accretion disks in high-mass protostars appears to be clear [109, 158, 182], but the definition of "disk" may need to be relaxed (see Cesaroni et al. [45], and references therein, see also Fig. 1.3). If massive protostellar disks are rotationally supported and stable they should present rotation profiles close to Keplerian

$$V_{rot}(r) \propto r^{-1/2}, \tag{1.6}$$

as has been found in a few disks around high-mass protostars [45]. However, in most MSFRs, the rotating structures appear to have messier velocity fields, and infall motions appear to be $V_{inf} \sim V_{rot}$, which suggests that these disks are not rotationally-supported, flat structures, but rather spiraling-in "toroids". This may well be caused by angular resolution limitations[2] and contamination from the dense envelope in the selected dust/gas tracer. The initial angular momentum of the collapsing envelope is also important in determining the radius at which the rotating structure becomes a rotationally supported disk [106]. Therefore, the aspect ratio, size, and velocity field of an envelope + disk that is not fully resolved is biased by observational limitations (resolution and tracer) and the real physical conditions of the collapsing core.

---

[2]The highest angular resolution achieved by current mm interferometers is $\sim$0.5″, or 5,000 AU at a distance of 5 kpc.

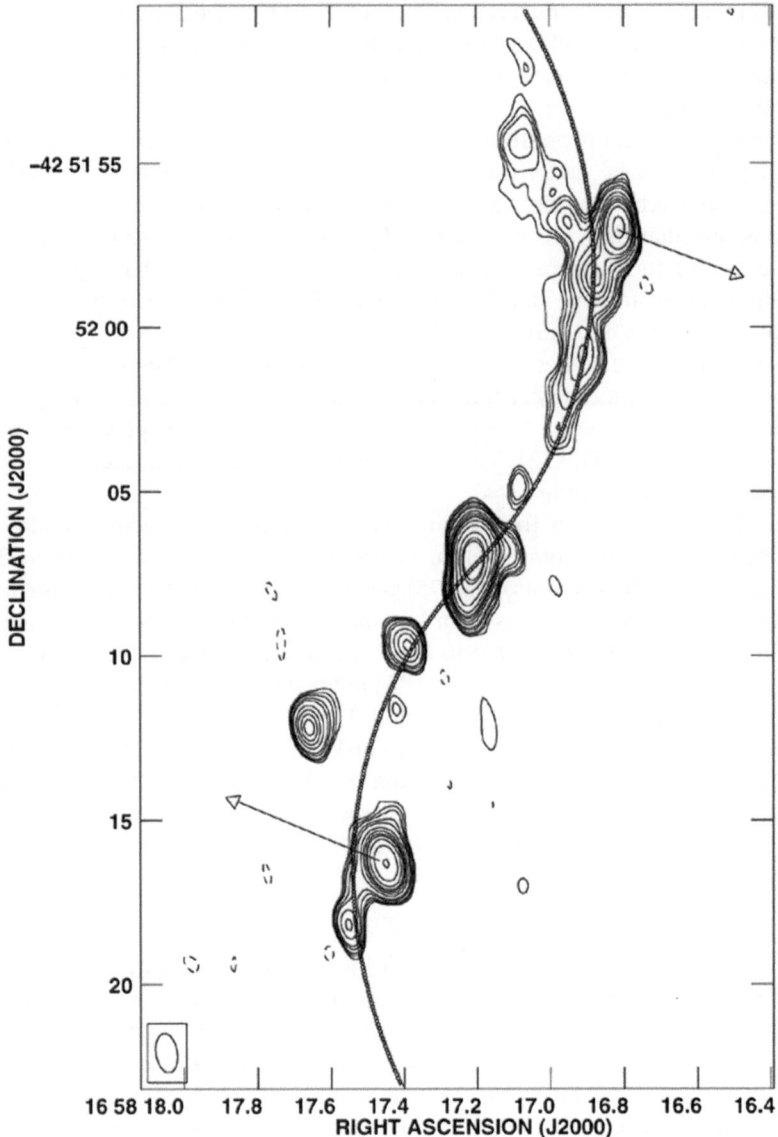

Fig. 1.5 Free-free radio jet in the high-mass protostar IRAS 16547−4247. The arc-shaped lobes may be due to precession of the disk-jet system. Credit Rodríguez et al. [175] (Reproduced by permission of the AAS)

Another aspect that appears to be key in massive protostellar disks is their gravitational (in)stability. Equation 1.6 is only valid if the disk mass is much smaller than the stellar mass $M_{disk} \ll M_\star$. However, it has been found that the masses of the rotating structures that surround high-mass stars are comparable or even much larger than the stellar mass. Still, it is difficult to determine what fraction of this mass corresponds to the true disk (see above). Recent models and numerical simulations of MSF indicate that these disks are unstable and prone to fragment due to their own gravity [120, 122, 163], which actually helps to explain the higher fraction of multiple systems for high-mass stars [127].

To avoid confusing the discussion during the rest of this thesis, in most times we will use the term "accretion flow" to refer to the rotating/infalling structures around high-mass protostars.

1.3   The Formation of the Most Massive Stars

1.3.1   From Massive to Very Massive

Early calculations of star formation showed that for spherical accretion, the radiation pressure exerted by a massive protostar of about $M_\star \sim 8 M_\odot$ was enough to stop further accretion [102, 220], leading to an apparent contradiction with the observation that stars as massive as $M_\star \sim 100 M_\odot$ do exist. As mentioned in previous sections, both observations and theory have concurred in that geometry is not spherical: the presence of disks focuses radiation pressure in the polar directions and permits accretion in the disk plane [121, 224].

As the stellar mass keeps growing, a new ingredient may need to be taken into account: photoionization. The amount of UV photons produced by a star increases very rapidly with mass [203]. Therefore, while stars with masses in the range $M_\star \sim 5$ to $15 M_\odot$ are not able to photoionize enough hydrogen to detectable levels,[3] stars with $M_\star$ roughly greater than $15 M_\odot$ start to photoionize their own accretion flow [105]. It is to these O-type stars that in the rest of this thesis we refer to as "the most" or "very" massive stars.

1.3.2   Ultracompact and Hypercompact HII Regions

Observationally, the existence of a class of very small ($R < 0.1 \, pc$) and dense ($n > 10^4 \, cm^{-3}$) H II regions, the so-called ultracompact (UC) and hypercompact (HC) H II regions (see Table 1.2) has been known for two decades [57, 125, 221]. In the simplest interpretation, they are often thought to be freely expanding into their surrounding medium at the sound speed of the ionized gas $c_{HII} \sim 10 \, km \, s^{-1}$. Such expectation comes from extrapolating the knowledge that larger H II regions (from

---

[3]When observed, the ionization in these sources may arise from shock-induced ionization as in low-mass jets.

Table 1.2 Properties of HII regions[a]

Class	Size (pc)	Density (cm^{-3})	Ionized mass ($M_\odot$)
Hypercompact (HC)	<0.02	>10^5	~10^{-3}
Ultracompact (UC)	≲0.1	>10^4	~10^{-2}
Compact	≲0.5	≳10^3	~1
Classical	~10	~100	~10^5
Giant	~100	~10	10^3–10^6
Supergiant	>100	~10	10^6–10^8

[a] Adapted from Kurtz [124]. The separation between UC and HC HII regions is loosely defined. It may be better to use size to define them, since density and ionized-mass estimations are model dependent (see, e.g., Chap. 1)

compact to giant) do expand hydrodynamically due to the pressure contrast of the hot ionized gas ($T_{HII} \sim 10,000$ K) with the surrounding neutral gas ($T_{neut} \sim 10$ K) [193]. However, since their discovery, it has been known that there are too many UC and HC H II regions compared to the number expected if they are freely expanding [46, 125, 221]. Another piece of the puzzle is that UC and HC H II regions have varied morphologies: spherical/unresolved, cometary, core-halo, shell-like, irregular, and bipolar [57, 125, 221].

Many ideas have been put forward that partially explain the varied morphologies and long lifetimes of UC and HC H II regions. In the most promising analytical models, some of the bipolar and unresolved H II regions may result from the ionized gas expelled by accretion disks that are being photoevaporated [93, 137], while some of the cometary H II regions may be "champagne flows", i.e., ionized gas that is expelled from a neutral cloud in a preferential direction due to the presence of an initial density gradient [11, 198].

1.3.3 Accretion and Ionization

If the gravity of the star(s) is taken into account, HC H II regions can be gravitationally trapped [104, 105] when they have a radius smaller than the gravitational radius

$$R_g \approx GM_\star/c_{HII}^2, \tag{1.7}$$

where $c_{HII} \sim 10$ km s^{-1} is the speed of sound in the ionized gas. $R_g \sim 100$ AU for a star with mass $M_\star = 20 M_\odot$. In trapped H II regions, the accretion flows continue and gas accretes onto the star in spite of being ionized. Most of the well studied HC and UC H II regions have observed sizes larger than their gravitational radius, however this does not mean that they cannot harbor stars that are still accreting. Considering that geometry is unlikely to be spherical, it may be the case that in some solid angles the H II region has a radius $R_{HII} > R_g$ and therefore there is an ionized outflow, while in some other solid angles (e.g., the disk plane) $R_{HII} < R_g$ and accretion proceeds onto the star [106]. In this thesis we have observed MSFRs that may harbor very massive stars that are still accreting in spite of having a small H II region. We have found, both from observations (Chaps. 2–4) and from analyzing numerical simulations (Chap. 5) that this is a very feasible idea. Observations with the next generation of interferometers like the Expanded Very Large Array (EVLA) and the Atacama Large Millimeter/submillimeter Array (ALMA) are needed to test this model further, and to get to an evolutionary picture where both the observations of shock-ionized jets and photoionized regions can be understood within a unified framework.

1.4 Goals

It is the purpose of this thesis to *improve our understanding of the formation of the most massive (O-type) stars in our Galaxy*. To achieve this, we have made extensive use of the Submillimeter Array (SMA) in Mauna Kea, Hawaii, and of the Very Large Array (VLA) in Socorro, New Mexico. We have also compared our results with analytical models and state-of-the-art numerical simulations. Two important questions that need to be answered to achieve our goal are:

- Does photoionization stop accretion? or does mass growth continue after the onset of an H II region?
- What is the relation of the ionized and molecular gas from disk ($\sim$1,000 AU) scales to core (0.1 pc) to clump (1 pc) scales?

The next four chapters describe the main results of this thesis. Chapters 2 and 3 are case studies of two very luminous $L > 10^5 L_\odot$ MSFRs. The first one, W33A has only very faint free-free emission, indicating that ionization has just started. The second one, G20.08N, has a more developed cluster of UC and HC H II regions at the center. Both regions have indications of active accretion at disk and core scales. Also, both regions have indications of large-scale (pc) collapse and converging motions that indicate that the clump-scale gas is actively participating in star formation. Chapter 4 reports the detection of a flux decrement in the HC H II region G24.78 A1. This results indicates that the H II region contracted in a period of several years, a result inconsistent with the simple expectation of ever expanding H II regions. Rather, we interpret this result as evidence for density changes in the inner (ionized) accretion flow and active accretion in the HC H II region stage. Chapter 5 is an analysis of the properties as a function of time of the H II regions in the cluster formation simulation recently presented by Peters et al. [161]. Variability and size changes (including negative variations) are a natural result in these simulations due to the constant evolution of the partially-ionized accretion flow. We quantify the variability in the simulations and give predictions for future surveys looking for variability.

Chapter 2
A MSFR at the Onset of Ionization: W33A

2.1 Summary

Interferometric observations of the W33A massive star-formation region, using the Submillimeter Array (SMA) and the Very Large Array (VLA) at resolutions from $5''$ ($\approx 0.1\,pc$) to $0.5''$ (≈ 0.01 pc) are presented. Our main findings are: (1) we detected parsec-scale filaments of cold molecular gas. Two filaments at different velocities intersect in the zone where the star formation is occurring, consistent with triggering of the star-formation activity by the convergence of such filaments. This has been predicted by numerical simulations of star formation initiated by converging flows. (2) The two dusty cores (MM1 and MM2) at the intersection of the filaments are found to be at different evolutionary stages. Each of them is resolved into multiple condensations. MM1 and MM2 have different temperatures, continuum spectral indices, molecular-line spectra, and masses of both stars and gas. (3) The dynamics of the "hot-core" MM1 indicates the presence of a rotating disk in its center (MM1-Main) around a faint peak of ionized-gas emission. The stellar mass is estimated to be $\sim 10 M_{\odot}$. A massive molecular outflow is observed to emanate perpendicular to the disk.

These results have been published in Galván-Madrid, Roberto, Zhang, Qizhou, Keto, Eric R., Ho Paul T. P., Zapata, Luis A., Rodríguez, Luis F., Pineda, Jaime E., and Vázquez-Semadeni, Enrique. "From the Convergence of Filaments to Disk-Outflow Accretion: Massive-Star Formation in W33A". The Astrophysical Journal, 725, 17 (December 2010) [75].

2.2 Introduction

Stars form by accretion of gas in dense molecular-cloud cores. However, the differences, if any, in the details of the formation process of massive stars (those with

R.J. Galván-Madrid, *On the Formation of the Most Massive Stars in the Galaxy*,   15
Springer Theses, DOI 10.1007/978-1-4614-3308-8_2,
© Springer Science+Business Media New York 2012

roughly $M_\star > 8 M_\odot$) compared to low-mass stars are not well understood. Recent reviews on the topic are those by Beuther et al. [30] and Zinnecker and Yorke [238].

Our program is aimed at studying how the formation of massive stars in clusters proceeds in the presence of different levels of ionization, from the onset of detectable free–free emission to the presence of several bright ultracompact (UC) H II regions. In this chapter we present our first results on the massive star-formation region W33A (also known as G12.91-0.26), at a kinematic distance of 3.8 kpc [96]. W33A is part of the W33 giant H II region complex [214]. It was recognized as a region with very high far infrared luminosity ($\approx 1 \times 10^5 L_\odot$), but very faint radio-continuum emission by Stier et al. [196]. van der Tak et al. [205] modeled the large-scale (arcminute) cloud as a spherical envelope with a power-law density gradient, based on single-dish mm/submm observations. Those authors also presented mm interferometric observations at several-arcsecond resolution that resolved the central region into two dusty cores separated by $\sim$20,000 AU. The brightest mm core contains faint ($\sim$1 mJy at cm wavelengths) radio-continuum emission [169] resolved at 7 mm into possibly three sources separated by less than $1''$ ($\approx$4,000 AU) from each other [204]. These radio sources were interpreted by van der Tak and Menten [204] as the gravitationally trapped H II regions set forth by Keto [105]. However, the earlier detection by Bunn et al. [36] of near-infrared recombination line (Brα) emission with FWHM = 155 km s^{-1} suggests that at least some of the radio free–free emission is produced by a fast ionized outflow. More recently, Davies et al. [55] reported spectroastrometry observations of Brγ emission toward W33A. The Brγ emission appears to be produced by at least two physical components: broad line wings extending to a few hundreds of kilometers per second from the systemic velocity appear to trace a bipolar jet on scales of a few AU, while the narrow-line emission may be attributed to a dense H II region [55]. Being a bright mid- and far-infrared source, W33A has also been target of interferometry experiments at these wavelengths, which reveal density gradients and non-spherical geometry in the warm dust within the inner few hundred AU [58,59].

In this chapter we report on millimeter and centimeter interferometric observations performed with the Submillimeter Array (SMA) and the Very Large Array (VLA) at angular resolutions from $\sim5''$ to $0.5''$. We find a massive star-forming cluster embedded in a parsec-scale filamentary structure of cold molecular gas. The dense gas is hierarchically fragmented into two main dusty cores, each of them resolved into more peaks at our highest angular resolution. The main cores appear to be at different evolutionary stages, as evidenced from their differing spectra, masses, temperatures, and continuum spectral indices. The warmer core harbors faint free–free emission centered on a rotating disk traced by warm molecular gas. The disk powers a massive molecular outflow, indicating active accretion. In Sect. 2.3 of this chapter, we describe the observational setup. Section 2.4 we list our results, in Sect. 2.5 we present a discussion of our findings, and in Sect. 2.6 we give our conclusions.

2.3 Observations

2.3.1 SMA

We observed the W33A region with the Submillimeter Array[1] in the 1.3-mm (230 GHz) band using two different array configurations. Compact-array observations were taken on 2007 July 17, and covered baselines with lengths between 7 and 100 Kλ (detecting spatial structures in the range of 29.5–2.1''). Very Extended (VEX) configuration data were taken on 2008 August 2, with baseline lengths from 23 to 391 Kλ (9.0–0.5''). For both observations, the two sidebands covered the frequency ranges of 219.3–221.3 and 229.3–231.3 GHz with a uniform spectral resolution of $\approx$0.5 km s^{-1}.

We also report on the continuum emission from archival observations taken in the 0.9-mm (336 GHz) band on 2006 May 22. The array was in its Extended configuration, with baseline lengths from 18 to 232 Kλ (11.4–0.8''). These data were used to constrain the spectral index of the continuum sources.

The visibilities of each data set were separately calibrated using the SMA's data calibration program, MIR. We used Callisto to obtain the absolute amplitude and quasars to derive the time-dependent phase corrections and frequency-dependent bandpass corrections. Table 2.1 lists relevant information on the calibrators. We estimate our flux-scale uncertainty to be better than 15%. Further imaging and processing was done in MIRIAD and AIPS.

The continuum was constructed in the (u, v) domain from the line-free channels. The line-free continuum in the 1.3-mm Compact-configuration data was bright enough to perform phase self-calibration. The derived gain corrections were applied to the respective line data. No self-calibration was done for the higher angular resolution data sets.

Table 2.1 Observational parameters

Epoch	Array	Phase center[a] α(J2000)	δ(J2000)	Bandpass calibrator	Flux calibrator	Phase calibrator
2004 06 14+15	VLA-D	18 14 39.500	−17 51 59.800	3C273	3C286	1,851 + 005
2006 05 22	SMA-Extended	18 14 39.509	−17 51 59.999	3C273	Callisto	1,733 − 130
2007 07 17	SMA-Compact	18 14 39.495	−17 51 59.800	3C454.3	Callisto	1,733 − 130
2008 08 02	SMA-VEX	18 14 39.495	−17 51 59.800	3C454.3	Callisto	1,733 − 130

[a] Units of R.A. are hours, minutes, and seconds. Units of decl. are degrees, arcminutes, and arcseconds

---

[1] The Submillimeter Array is a joint project between the Smithsonian Astrophysical Observatory and the Academia Sinica Institute of Astronomy and Astrophysics and is funded by the Smithsonian Institution and the Academia Sinica. See Ho et al. [89].

2.3.2   VLA

We observed the $(J,K) = (1,1)$ and $(2,2)$ inversion transitions of NH_3 with the Very Large Array.[2] Observations were carried out on 2004 June 14 and 15 (project AC733). The array was in its D configuration, with baseline lengths in the range of 3–79 $K\lambda$ (detecting scales from 68.7″ to 2.6″). The correlator was set to the 4-IF mode. Each of the IF pairs was tuned to the (1,1) and (2,2) lines, respectively, covering a bandwidth of 3.1 MHz (39 km s^{-1}) at a spectral resolution of 0.6 km s^{-1}.

The data were calibrated and imaged using standard procedures in the AIPS software. Table 2.1 lists the quasars used to derive the absolute flux scale, the time-dependent gain corrections, and the frequency-dependent passband calibration. The absolute flux scale is accurate within a few percent. No self-calibration was performed.

2.4   Results

2.4.1   Continuum Emission

2.4.1.1   Morphology

Our observations at 1.3 mm resolve each of the 2 mm cores reported by [205] into multiple continuum sources. The concatenation of Compact- and VEX-configuration data permits us to simultaneously resolve the structures at $\approx 0.5″$ resolution and to be sensitive to relatively extended structures. Figure 2.1 (*left*) shows the 1.3-mm continuum map. It is seen that MM1 and MM2 are resolved into at least three and two smaller mm peaks respectively (marked by crosses in Fig. 2.1). Only MM1 is associated with the cm emission detected by [169]. Two of the three faint 7-mm sources ($S_{7\,mm} \sim 1$ mJy) reported by [204] toward MM1 at a resolution of $\sim 0.05″$ (marked by triangles in Fig. 2.1) are counterparts of the 1.3-mm peaks. The faintest 7-mm source has no association in our continuum or line data. In addition to the clearly identified 1.3-mm peaks, the northeast-southwest large-scale continuum ridge appears to have more fainter sources. Another possible source is well separated from the ridge, at $\approx 8″$ to the southwest of MM1. More sensitive observations are necessary to investigate their nature.

We label the identified mm peaks as MM1-Main (the brightest source of MM1), MM1-NW (for northwest), MM1-SE (for southeast), MM2-Main, and MM2-NE. Table 2.2 lists the peak positions and peak intensities measured in the mm map

---

[2]The National Radio Astronomy Observatory is operated by Associated Universities, Inc., under cooperative agreement with the National Science Foundation.

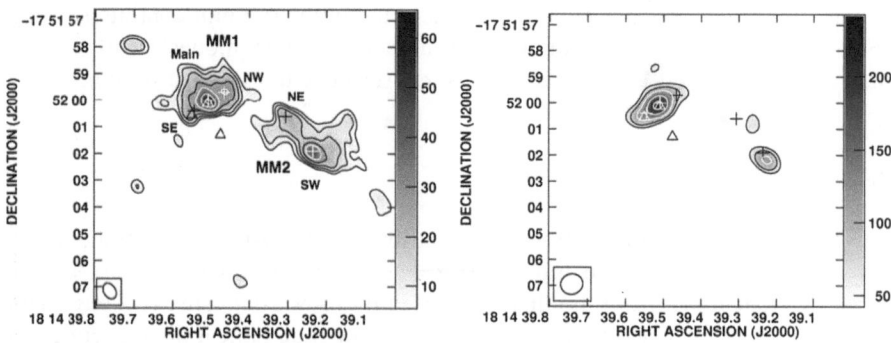

Fig. 2.1 (Sub)millimeter continuum emission in W33A. The *left panel* shows the 231 GHz (1.3 mm) continuum from the SMA Compact+VEX data (HPBW = 0.63″ × 0.43″, *P.A.* = 30.7°). Contours are at −5, 5, 7, 10, 15, 20, 30, and 40 times the noise of 1.5 mJy beam^{-1}. The *right panel* shows the 336 GHz (0.9 mm) continuum from the Extended-configuration data (HPBW = 0.88″ × 0.83″, *P.A.* = 275.1°), with contours at −5, 5, 7, 10, 15, 20, 30, and 39 times the rms noise of 6 mJy beam^{-1}. The cores MM1 and MM2 are labeled, and the sources into which they fragment are marked by *crosses*. *Triangles* mark the positions of the faint 7 mm sources reported by van der Tak and Menten [204]. One arcsec corresponds to 3,800 AU (0.018 pc)

Table 2.2 Millimeter continuum sources

Core[a]	Component[b]	α(J2000)[c] (hr, min, s)	δ(J2000)[c] (deg, arcmin, arcsec)	I_{peak}(1.3 mm)[d] (mjyb)	S(1.3 mm)[e] (mJy)
MM1	MM1-NW	18 14 39.47	−17 51 59.7	31	357
	MM1-Main	18 14 39.51	−17 52 00.0	65	357
	MM1-SE	18 14 39.55	−17 52 00.4	25	357
MM2	MM2-Main	18 14 39.24	−17 52 01.9	43	289
	MM2-NE	18 14 39.31	−17 52 00.6	22	289

[a] Main core as labeled in Fig. 2.1, left panel.
[b] Clearly distinct subcomponents of the main cores as marked in Fig. 2.1.
[c] Position of peak.
[d] Peak intensity ±1.5 mJy beam^{-1}. HPBW = 0.63″ × 0.43″.
[e] Added flux of the subcomponents of each core. The uncertainties in the fluxes of MM1 and MM2 are ±20 mJy and ±25 mJy respectively.

of Fig. 2.1 *left*. The sums of the 1.3-mm fluxes of the components that we obtain from multi-component Gaussian fits to the sources comprising MM1 and MM2 are robust, and consistent with the fluxes measured by integrating the intensity over the areas of interest. However, the sizes and fluxes of the individual components in the fits are not accurate, mainly because of insufficient angular resolution. Table 2.2 lists the added flux of the subcomponents of MM1 and MM2. The ratio of the 1.3-mm flux of MM1 to that of MM2 in our data is 1.2, very close to that reported by [205]: 1.3. The fluxes that we report are 88–106% larger than those in [205], probably due to differences in (u, v) coverage and flux-scale uncertainties.

Only the bright, compact sources are detected in the 0.9-mm continuum image (Fig. 2.1 *right*). This single-configuration data set has a more modest (u, v) coverage than the concatenated 1.3-mm data and is less sensitive to extended structures.

2.4.1.2   Nature of the Continuum

To set an upper limit to the free–free contribution at 1.3-mm we extrapolate the 8.4–43.3-GHz free–free spectral index $\alpha = 1.03 \pm 0.08$ (where the flux goes as $S_\nu \propto \nu^\alpha$), calculated from the fluxes reported by [169, 204]. This is a reasonable assumption since α for free–free sources with moderate optical depths, either jets (e.g., [35, 79, 92]) or H II regions (e.g., [66, 74, 114]), is approximately in the range from 0.5 to 1.

For MM1 with a 7-mm flux of $S_{7\,\mathrm{mm}} \approx 4.3\,\mathrm{mJy}$, the maximum free–free flux at 1.3-mm is 28 mJy. The 1.3-mm flux integrated over MM1 is $S_{1.3\,\mathrm{mm}} \approx 357 \pm 25\,\mathrm{mJy}$ (Table 2.2). Therefore, the free–free contribution to the 1.3-mm flux is at most $\sim$8%. No cm continuum has been detected toward MM2 or in the rest of the field, thus, the 1.3-mm emission outside MM1 is most probably produced entirely by dust. Using the same considerations, the free–free flux of MM1 at 0.9 mm is less than 42 mJy. The integrated flux of MM1 at this wavelength is $S_{0.9\,\mathrm{mm}} \sim 612\,\mathrm{mJy}$, then the free–free emission is at most $\sim$7%. Because the data are taken at different epochs, the possibility of radio variability (see [67, 73, 206], for reports on other targets) adds to the uncertainty. The 0.9-mm data may suffer from missing flux, making the fractional free–free contribution at this wavelength even smaller. In the rest of this study, we consider the (sub)mm free–free emission to be negligible.

To compare the 0.9- and 1.3-mm fluxes in a consistent way, we produced images with a uniform (u, v) coverage (using only baselines with lengths from 30 to 230 Kλ), without self-calibration (using only the VEX data at 1.3 mm and the Extended data at 0.9 mm), and a common circular synthesized beam (HPBW $= 0.85''$). The average spectral indices of the two main mm cores are $\langle \alpha_{\mathrm{MM1}} \rangle = 3.3 \pm 0.3$ and $\langle \alpha_{\mathrm{MM2}} \rangle = 2.5 \pm 0.4$. In the Rayleigh–Jeans (R-J) approximation ($h\nu \ll k_B T$), the spectral index of thermal dust emission is $\alpha = 2 + \beta$, where β is the exponent of the dust absorption coefficient. The fiducial interstellar–medium (ISM) value of β is 2, while for hot cores in massive star-forming regions (MSFRs) typical values are $\beta \approx 1 - 2$ [43, 235]. Therefore, MM1 has $\beta \approx 1.3$ typical of a hot core, but MM2 has $\beta \approx 0.5$. In Sect. 2.4.2 we show that the kinetic temperature of MM2 is $\approx$46 K, then the R-J limit is not a good approximation at 0.9 mm for MM2.

Without using the assumption of being in the R-J limit, the gas mass M_{gas} derived from optically–thin dust emission at 1.3-mm can be obtained from Kirchhoff's law:

$$\left[\frac{M_{\mathrm{gas}}}{M_\odot}\right] = (26.6) \times \left(\exp\left(\frac{11.1}{[T_{\mathrm{dust}}/\mathrm{K}]} \right) - 1 \right) \times \left(\frac{[F_{1.3\,\mathrm{mm}}/\mathrm{Jy}][d/\mathrm{kpc}]^2}{[\kappa_{1.3\,\mathrm{mm}}/\mathrm{cm}^2\mathrm{g}^{-1}]} \right), \quad (2.1)$$

where T_{dust} is the dust temperature, $F_{1.3\,\text{mm}}$ is the 1.3-mm flux density, d is the distance to the object, and $\kappa_{1.3\,\text{mm}}$ is the dust absorption coefficient. Assuming coupling between gas and dust, the dust temperature in MM1 is ≈ 347 K in the inner arcsecond (obtained from fits to CH_3CN lines, see Sect. 2.4.2) and > 100 K at larger scales (obtained from NH_3 lines, see Sect. 2.4.2). For this range of temperature, using an opacity $\kappa_{1.3\,\text{mm}} = 0.5\,\text{cm}^2\,\text{g}^{-1}$ [156], (2.1) gives a mass for MM1 in the range $M_{\text{MM1}} = [9, 32]\,M_{\odot}$. For MM2 with a temperature of 46 K (Sect. 2.4.2), the mass is $M_{\text{MM2}} \sim 60\,M_{\odot}$. MM2 then appears to be much colder and more massive (in gas) than MM1. The uncertainties in opacity make the mass estimation accurate to only within a factor of a few.

2.4.2 Molecular Line Emission

2.4.2.1 The Parsec-Scale Gas

The large-scale gas within an area of $\sim 1'' \times 1''$ (or ~ 1 pc) can be divided into quiescent gas and high-velocity gas. The quiescent gas is best traced by the VLA NH_3 data. The high-velocity gas is seen in the SMA CO (2–1) maps.

2.4.2.2 Morphology and Velocity Structure

There is a clear morphological difference between the quiescent and the high-velocity gas. Figure 2.2 shows moment maps of the NH_3 (2,2) line overlaid with the high-velocity CO gas. The NH_3 moment maps were integrated in the $[31, 43]$ km s^{-1} LSR velocity range. The blueshifted CO gas was integrated in the range $[0, 22]$ km s^{-1}, and the redshifted CO was integrated in $[62, 98]$ km s^{-1}. The systemic velocity of the gas closer to MM1 is $V_{\text{sys}} \approx 38.5$ km s^{-1} (Sect. 2.4.2). The quiescent NH_3 emission is composed of one prominent filamentary structure in the east-west direction that peaks toward the MM1 region (Fig. 2.2), plus another filamentary structure that extends to the south of MM1 and MM2, and some fainter clumps toward the northwest of MM2. The high-velocity CO traces at least two molecular outflows that expand outward off the quiescent filaments. The lobes of the most prominent outflow are centered in MM1, and extend toward the northwest (redshifted gas) and southeast (blueshifted) at a position angle P.A. $\approx 133°$. The observed size of this outflow is about 0.4 pc. The redshifted lobe of a second high-velocity outflow extends ≈ 0.5 pc to the north-northeast of the cores at P.A. $\approx 19°$, and appears to be originated in MM2. The blueshifted side of this outflow does not appear at high velocities. The P.A. that we find for the main outflow agrees very well with the P.A. $\approx 135°$ of the outflow as seen at 2.2 µm reported by Davies et al. [55, see their Fig. 1], which matches an elongated 4.5 µm structure in *Spitzer* images (Figure 12 of [49]). However, the infrared emission is three to four times larger. Also, using single-dish observations, [59] reported a CO $J = 3 - 2$ outflow whose

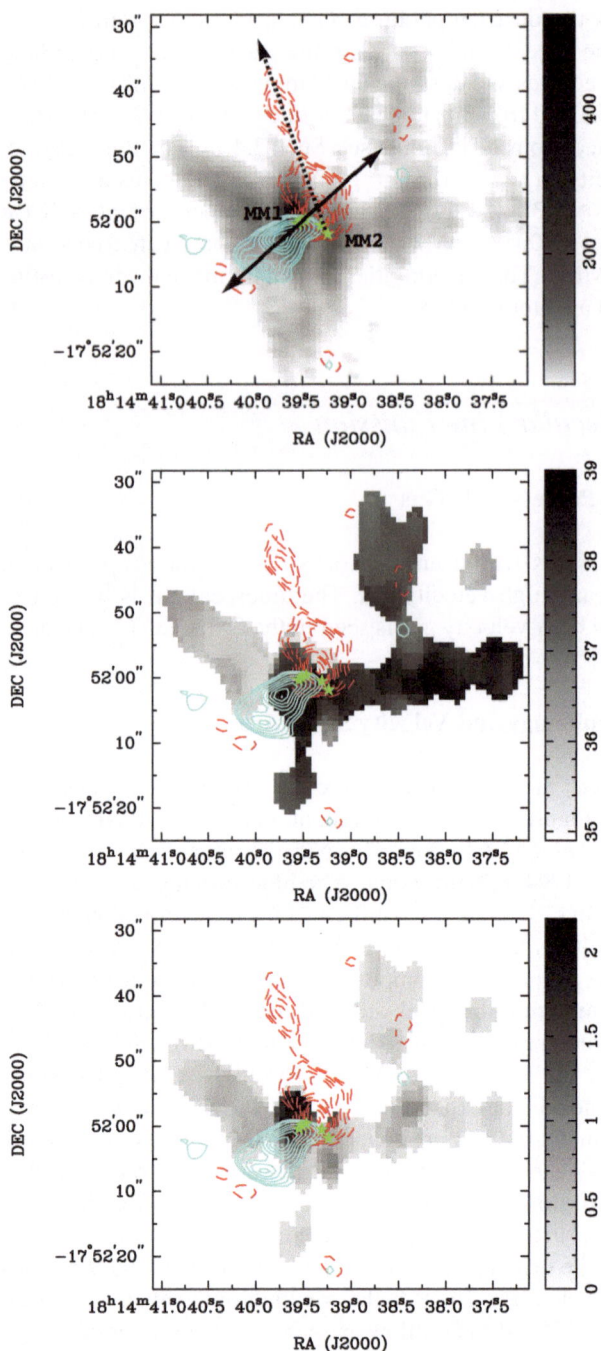

Fig. 2.2 (continued)

orientation matches those of both the SMA and the near IR outflows. The low-velocity CO (2–1) gas could not be properly imaged because of the lack of short (u,v) spacings.[3]

The large-scale quiescent gas has two velocity components, which appear to be two different structures of gas at different velocities, as can be seen in Fig. 2.2. The gas associated with MM1 and MM2 (center of the main filament), as well as the western part of the main filament and the north–south extensions appears to be at $\approx$38.5 km s^{-1}, with a typical mean-velocity dispersion of 0.4 km s^{-1}. The eastern part of the main filament appears blueshifted, with a mean centroid velocity of $\approx$35.9 km s^{-1} and a dispersion of the centroid velocity of 0.4 km s^{-1}. The two gas structures overlap in space toward the MM1/MM2 region, which suggests that the star formation activity in these mm cores was triggered by the convergence of the filaments of molecular gas. Figure 2.2 shows a three-dimensional rendering of the same data that better illustrates this result. At the position of the mm cores, the filamentary structures do not merely superpose in position-position-velocity space, but merge into a region that suddenly extends to higher velocities. The larger velocity range at the center (see also Fig. 2.2) is found to be due to coherent velocity structures (disk and outflows in the dense gas) with the SMA data (Sect. 2.4.2).

2.4.2.3 Physical Parameters

Now we derive the temperature structure of the parsec-scale filaments and lower limits to the outflow parameters.

Most of the gas in the pc-scale filaments, including the gas associated with MM2, is cold, with a kinetic temperature $T_{kin} = [20,50]$ K. T_{kin} rises significantly only toward MM1. A temperature map at the resolution of the NH$_3$ data (Fig. 2.3) was obtained by fitting the (1,1) and (2,2) line profiles as described in [177]. The errors in

---

Fig. 2.2 Parsec-scale gas structure toward W33A. The *color scale* shows the quiescent gas components as detected in the NH$_3$ (2,2) data (*HPBW* = 6.0″ × 2.6″, *PA* = 2°). The *blue solid contours* show the high-velocity gas as detected in CO (2–1) (*HPBW* = 3.0″ × 2.0″, *PA* = 56°) integrated in the range [0,22] km s^{-1}. The *red dashed contours* show the high-velocity CO gas in the range [62, 98] km s^{-1}. Contour levels are $-5, 5, 7, 10, 15, 20, 25, 30, 35, 40, 50$, and 60×0.7 Jy beam^{-1} km s^{-1}. The mm continuum sources identified in this study are marked with stars. The *top frame* shows the integrated NH$_3$ (2,2) intensity (moment 0, in mJy beam^{-1} km s^{-1}). The *middle panel* shows the intensity-weighted mean velocity (moment 1, in km s^{-1} with respect to the LSR). The *bottom panel* shows the velocity dispersion with respect to the mean velocity (moment 2, FWHM/2.35 in km s^{-1}). The direction of the identified outflows are marked with *arrows* in the *top panel*. Ten arcsec correspond to 38,000 AU (0.184 pc)

---

[3]Single-dish data were taken for this purpose but these were corrupted due to a bad off position and could not be used.

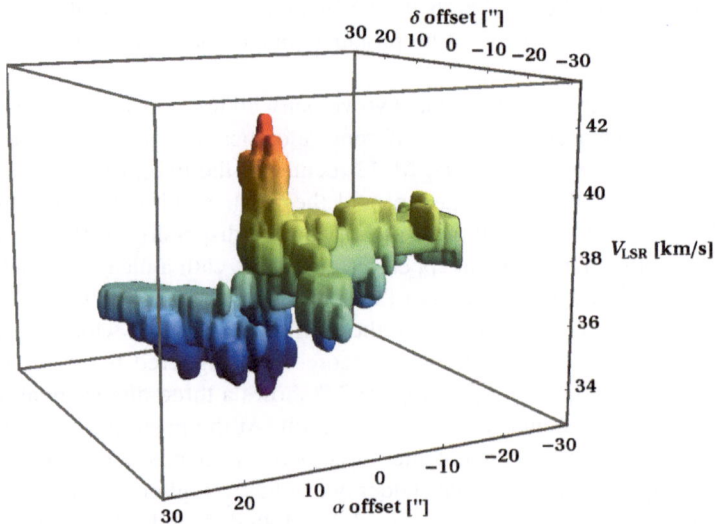

Fig. 2.3 Map of the kinetic temperature T_{kin} at large scales obtained from the NH$_3$ (1,1) and (2,2) data. It is seen that the pc-scale filaments are cold, with $T_{kin} = [20, 50]$ K. Only toward MM1 T_{kin} rises significantly, but the errors toward this region increase up to ≈ 40 K. Symbols are as in Fig. 2.1

the temperature determination are in general ~ 3 K, but get too large toward MM1.[4] We determine the temperature of MM2 to be $T_{MM2} \approx 46$ K. The temperature of MM1 is constrained to $T_{MM1} > 100$ K.

For the CO (2–1) line, the interferometric data suffer from missing flux for the more extended emission close to V_{sys}. We set the following limits to the outflow parameters: mass $M_{out} > 27 M_\odot$, momentum $P_{out} > 233 M_\odot$ km s^{-1}, and kinetic energy $E_{out} > 3 \times 10^{46}$ erg s^{-1}, where we corrected for the optical depth at each velocity bin using the ^{13}CO (2–1) line. We refer the reader to [167] for a description of the procedure to calculate the aforementioned quantities. The momentum and especially the energy estimations are less affected by missing flux since they depend more on the high-velocity channels. de Wit et al. [59] estimated the inclination angle of the inner-cavity walls of the outflow to be $i \sim 50°$, by radiative-transfer modeling of the mid-IR emission. Correcting by inclination, the lower limits to the momentum and energy of the outflow are $P_{out} > 362 M_\odot$ km s^{-1} and $E_{out} > 7 \times 10^{46}$ erg s^{-1}.

---

[4]The (2,2) to (1,1) ratio is not sensitive to temperatures much larger than 50 K, but we confirm the large temperatures in MM1 at smaller scales using the CH$_3$CN lines, see Sect. 2.4.2. The errors in the fits also increase toward MM1 due to its wider velocity structure.

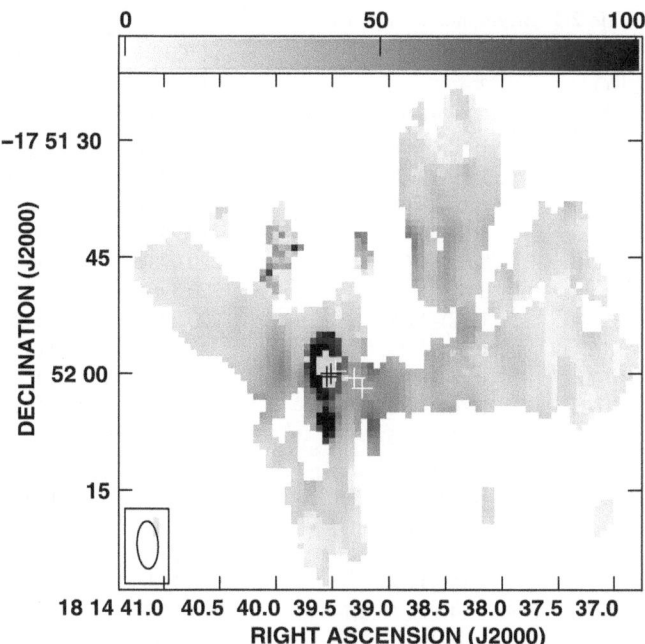

Fig. 2.4 Wide-band, continuum-free spectra in the image domain from the SMA Compact+VEX data at 1.3 mm. The *top row* shows the spectra for MM1 (the peak MM1-Main). The *bottom row* shows the spectra for MM2 (the peak MM2-Main). There is a striking difference in the richness of the spectra between the two cores. MM2 is almost devoid of molecular-line emission, in spite of it having a larger gas reservoir than MM1 (Sect. 2.4.1)

2.4.2.4 The Inner 0.1 pc

Morphology and Velocity Structure

The SMA data permit us to study the molecular gas at a resolution of $\approx 1,500$ AU ($0.4''$). Some molecular lines trace relatively cold gas, while some other lines trace the warmer gas closer to the heating sources. Figure 2.4 shows the spectra over the entire sidebands from the pixels at the positions of the mm peaks MM1-Main and MM2-Main. The prominent lines are labeled, and listed along with their upper-level energy in Table 2.3. Lines with a peak intensity below 20 K are not listed. A complete inventory of the molecular lines in W33A will be presented in the future. It is immediately seen that MM1 has a "hot-core" spectrum, while MM2 is almost devoid of molecular emission, if not for the CO, ^{13}CO and $C^{18}O$ $J = 2 - 1$, and faint SO $J(K) = 6(5) - 5(4)$ emission. We interpret this difference as a signature of the evolutionary stage of the cores, MM1 being more evolved than MM2.

Figure 2.5 shows the moment maps for three lines that exemplify what was mentioned above. The SO 6(5)–5(4) line (*top* row) extends in a ridge of ≈ 0.1 pc

Table 2.3 Bright molecular lines[a]

Species	Transition	ν_0 (GHz)	E_u (K)
$C^{18}O$	2–1	219.5603	15.8
HNCO	10(2,9)–9(2,8)	219.7338	228.4
–	10(2,8)–9(2,7)	219.7371	228.2
–	10(0,10)–9(0,9)	219.7982	58.0
$H_2{}^{13}CO$	3(1,2)–2(1,1)	219.9084	32.9
SO	6(5)–5(4)	219.9494	34.9
CH_3OH	8(0,8)–7(1,6) E	220.0784	96.6
^{13}CO	2–1	220.3986	15.86
CH_3CN	12(8)–11(8)	220.4758	525.5
–	12(7)–11(7)	220.5393	418.6
HNCO	10(1,9)–9(1,8)	220.5847	101.5
CH_3CN	12(6)–11(6)	220.5944	325.8
–	12(5)–11(5)	230.6410	247.3
–	12(4)–11(4)	220.6792	183.1
–	12(3)–11(3)	220.7090	133.1
–	12(2)–11(2)	220.7302	97.4
–	12(1)–11(1)	220.7430	76.0
–	12(0)–11(0)	220.7472	68.8
CH_3OH	15(4,11)–16(3,13) E	229.5890	374.4
–	8(−1,8)–7(0,7) E	229.7588	89.1
–	3(−2,2)–4(−1,4) E	230.0270	39.8
^{12}CO	2–1	230.5380	16.5
OCS	19–18	231.0609	110.8
^{13}CS	5–4	231.2207	26.6
CH_3OH	10(2,9)–9(3,6) $A-$	231.2811	165.3

[a]Molecular lines with peak $T_B \geq 20$ K. The first column refers to the molecule tag, the second column to the transition, the third to its rest frequency as found in Splatalogue (http://www.splatalogue.net/), and the fourth to the upper-level energy. Data used from Splatalogue are compiled from the CDMS catalog [151] and the NIST catalog [136]

long in the northeast–southwest direction, from MM1 to MM2. The emission is stronger toward MM1 and peaks in MM1-Main. Part of the emission toward MM2 is redshifted by ∼1–2 km s^{-1} with respect to the emission in the MM1 side, but there is no clear velocity pattern. Lines such as SO likely have large optical depths and trace only the surface of the emitting region, where clear velocity gradients, especially of rotation, may not be expected. From the SO data we constrain any velocity difference between the MM1 and MM2 cores to $\Delta V < 2$ km s^{-1}.

For a given molecule, the isotopologue lines and the lines with upper energy levels well above 100 K trace the more compact gas, closer to the heating sources. Figure 2.5 shows the examples of ^{13}CS $J = 5 - 4$ (*middle* panel) and CH_3CN $J(K) = 12(3) - 11(3)$ (*bottom*). Both of them are only visible toward MM1, and peak in MM1-Main. These lines trace a clear velocity gradient centered on MM1-Main, the blueshifted emission is toward the southwest, and the redshifted emission

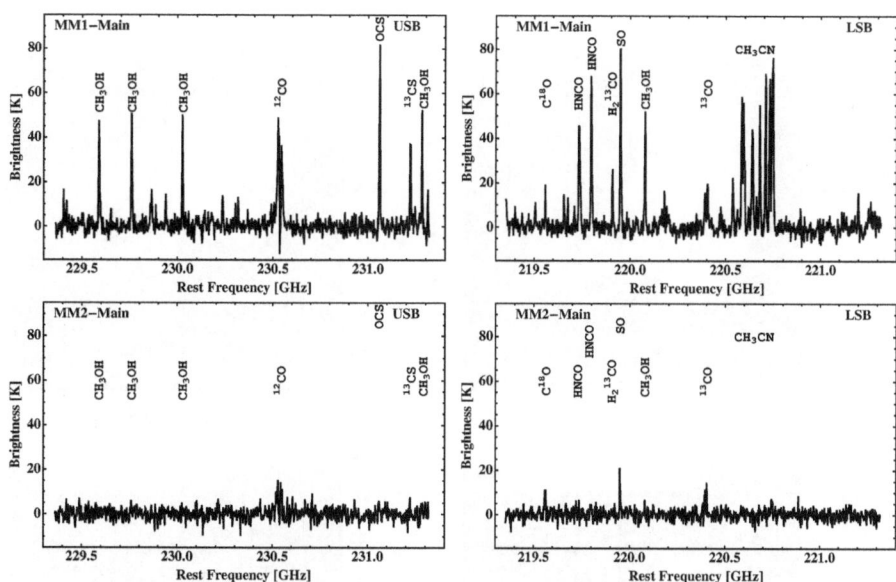

Fig. 2.5 Hot-core molecules toward the center of W33A. The *top row* shows the SO 6(5)–5(4) line. The *middle row* shows the ^{13}CS 5–4 line. The *bottom row* shows the CH$_3$CN 12(3)–11(3) line. *Contours* show the moment 0 maps at 5, 15, 30, 50, 100, 150, and 200 × 0.05 Jy beam^{-1} km s^{-1}. The *color scale* shows the moment 1 maps (*left column*) and moment 2 maps (*right column*). Symbols are as in Fig. 2.1. While the SO traces an extended envelope reaching MM2, the other molecules trace a clear velocity gradient indicative of rotation centered in MM1-Main. The velocity dispersion also peaks in MM1-Main

toward the northeast, perpendicular to the main bipolar CO outflow. We interpret this as rotation. The emission tracing this velocity gradient is not isolated, there is also emission coming from MM1-NW and MM1-SE. Especially in the CH$_3$CN lines, this extra emission appears to trace redshifted and blueshifted emission respectively. One possibility is that MM1-NW and MM2-SE are separate protostars from MM1-Main. However, the orientation of the lobes in the high-velocity outflow is the same. Therefore, we prefer the interpretation that MM1-NW and MM2-SE are not of protostellar nature, but emission enhancements (both in continuum and line emission) from the hot base of the powerful molecular outflow driven by the disk-like structure surrounding MM1-Main. In this scenario, the other 7-mm sources reported by [204] (or at least the counterpart of MM1-NW) can be interpreted as shocked free–free enhancements in a protostellar jet, similar to those observed in the high-mass star formation region IRAS 16547-4247 [68, 175].

Figure 2.6 shows the position–velocity (PV) diagrams of the CH$_3$CN $K = 3$ line shown in Fig. 2.5, centered at the position of MM1-Main perpendicular to the rotation axis (*top* frame) and along it (*bottom* frame). The rotation pattern is similar to those in objects that have been claimed to be Keplerian disks, i.e., structures where the mass of the central object is large compared to the mass of the

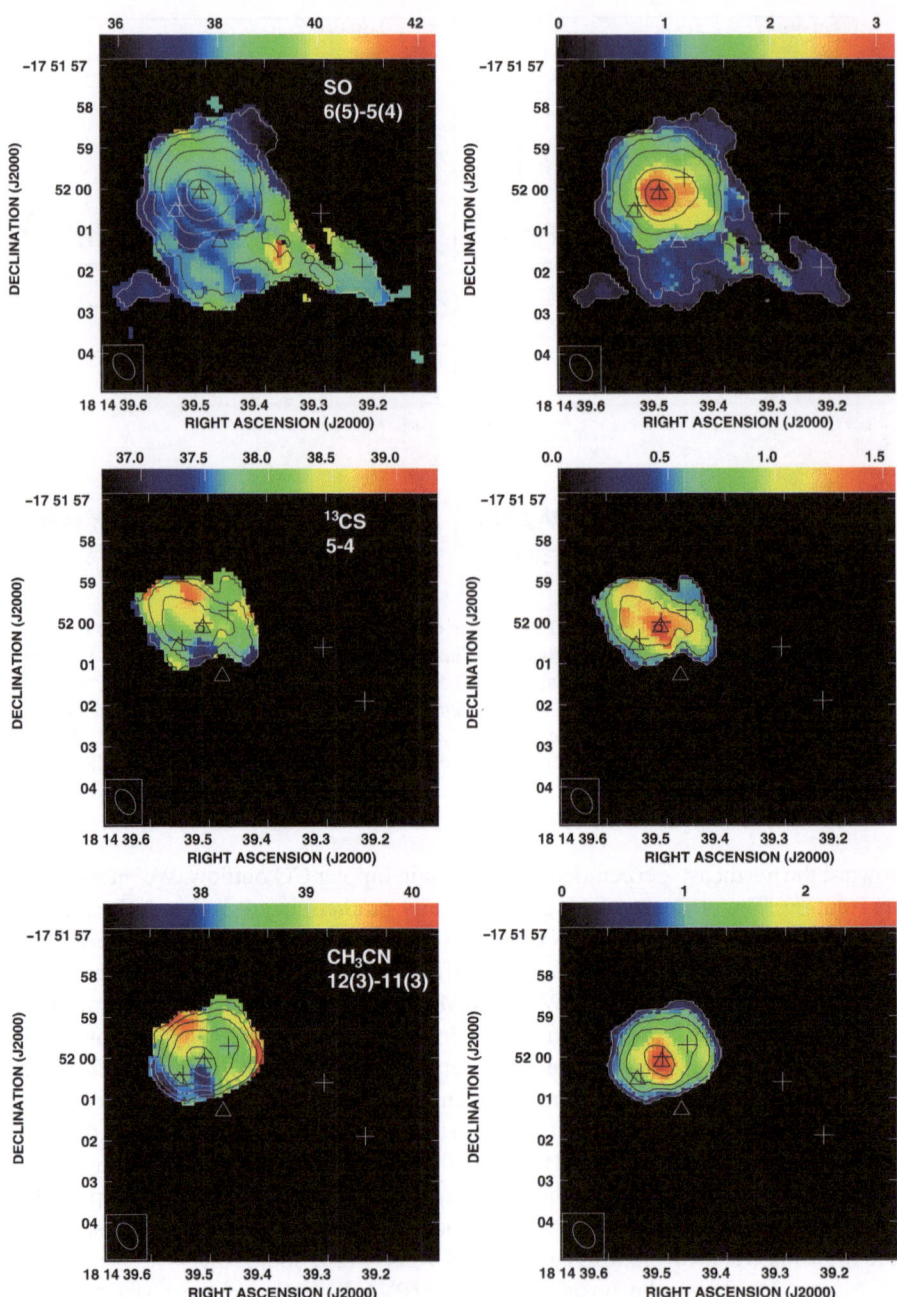

Fig. 2.6 (continued)

gas, rotating with a velocity $V_{rot} \propto r^{-0.5}$ [44, 98, 233]. The large velocity dispersion closest to MM1-Main (Figs. 2.5 and 2.6) ought to be caused by unresolved motions in the inner disk, since velocity dispersions well above 1 km s^{-1} cannot be due to the gas temperature.

Recently, [55] reported a possible disk-jet system centered in W33A MM1-Main. The jet, observed in the Brγ line, extends up to ± 300 km s^{-1} in velocity at scales ~ 1 AU, with a similar orientation and direction to the molecular outflow reported in this study. However, the velocity structure of what [55] interpret as a rotating disk has a similar orientation but opposite sense of rotation as the disk that we report. They used CO absorption lines with upper energy levels $E_u \sim 30$ K, while we use emission lines like those of CH$_3$CN, with $E_u > 100$ K (Table 2.3). If an extended screen of cold gas with a negligible velocity gradient is between the observer and the inner warm gas with a velocity gradient, it is possible that the absorption lines are partially filled with emission, mimicking a velocity gradient with the opposite sense to that seen in the emission lines.

2.4.2.5 Physical Parameters

Now we derive the temperature and column density of the hot-core emission, and constrain the stellar mass, gas mass, and CH$_3$CN abundance in MM1.

The kinetic temperature of the innermost gas can be obtained from the K lines of CH$_3$CN $J = 12 - 11$. To avoid the simplification of considering optically-thin emission assumed in a population-diagram analysis, we fit all the K lines assuming LTE, while simultaneously solving for the temperature T_{kin}, column density of CH$_3$CN molecules N_{CH_3CN}, and line width at half-power FWHM. The procedure to obtain the level populations can be found in [8].

Figure 2.7 shows the results of our fits to the CH$_3$CN spectra. The systemic velocity $V_{sys} \approx 38.5$ km s^{-1} was found to be optimal. The data outside the lines of interest have been suppressed for clarity, and the fit was done in the frequency windows where only the lines of interest are present. The gas is warmer (by 18%) and denser (by 415% in column) toward the peak MM1-Main (*bottom* frame) than in the average of the sources composing MM1 (*top* frame). This makes the case for the internal heating and a centrally-peaked density gradient in MM1-Main, as well as for its protostellar nature. Some lines are not completely well fit under the assumption of a single value for the parameters. On the one hand, the bright

---

Fig. 2.6 Position–Velocity (PV) diagrams for the CH$_3$CN $J = 12 - 11$ $K = 3$ data. The center is the position of MM1-Main (Table 2.2). *Top*: PV diagram at P.A. $= 39°$. A clear velocity gradient is seen from the southwest (blueshifted, negative position) to the northeast (redshifted, positive position). At the angle of this cut the velocity gradient has maximum symmetry. *Bottom*: PV diagram at P.A. $= 39° + 90° = 129°$. Negative positions are to the northwest, and positive positions to the southeast. There is emission in the range [36, 41] km s^{-1} at all positions. Closer to the center position, the velocity dispersion increases rapidly

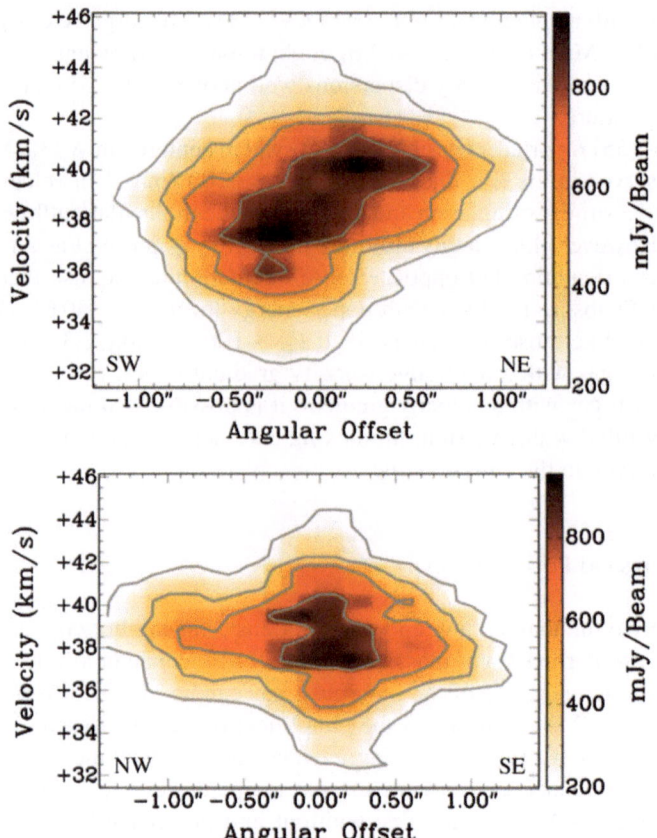

Fig. 2.7 Simultaneous fits to the CH_3CN $J = 12 - 11$ K lines. The dashed gray line is the data. The solid black line is the fit. The data outside the lines of interest have been set to zero to avoid contamination by other molecular lines. *Top:* Average spectra over the entire MM1 region. *Bottom:* Spectra toward the few central pixels at the peak position (MM1-Main). The gas is warmer, denser, and has a larger line width toward this position

$K = 7$ and 8 lines indicate the presence of some column of very warm gas; on the other hand, the $K = 3$ is not brighter than the $K < 3$ lines within the uncertainties, indicating that some column of gas is below 200 K. Two-component fits do not give better results. The reported values should be interpreted as an average along the line of sight. Detailed radiative transfer in the context of a physical model is currently under way.

Under the interpretation of edge-on rotation as the cause of the velocity gradient observed in Figs. 2.5 and 2.6, the dynamical mass in MM1-Main (stars plus gas) is about $9 M_\odot$, where a mean-velocity offset of 2.0 km s^{-1} with respect to $V_{sys} \approx$ 38.5 km s^{-1} was taken at a radius of 0.5$''$. Assuming that the disk is perpendicular to the outflow, and correcting for the outflow inclination angle $i = 50°$ estimated by [59], the enclosed dynamical mass is $M_{\mathrm{dyn}} \sim 15 M_\odot$.

Given that W33A is fragmented into multiple sources, a strict upper limit to the stellar mass in MM1-Main is $< 20 M_\odot$, the mass necessary to account for the total luminosity ($\sim 10^5 L_\odot$) of W33A. Also, the total gas mass in the MM1 sources inferred from dust emission is $\sim 10 M_\odot$, therefore the gas mass in the rotating structure around MM1-Main should be a fraction of it. For a gas mass in the rotating structure $M_{\mathrm{gas}} \sim 5 M_\odot$, the mass of the protostar(s) in MM1-Main amounts to $M_\star \sim 10 M_\odot$. MM1-Main appears to be B-type protostar still accreting from a rotating disk-like structure.

From the $\sim [9, 32] M_\odot$ gas-mass range in MM1 we derive an average column density of molecular gas of about $[1, 4] \times 10^{23}$ cm^{-2}. For an average CH$_3$CN column of $\sim 8 \times 10^{15}$ cm^{-2}, a CH$_3$CN abundance with respect to H$_2$ of $X(\mathrm{CH_3CN})$ $\sim [2, 8] \times 10^{-8}$ is derived, similar estimates in other regions of high-mass star formation [74, 168, 217].

2.5 Discussion

2.5.1 Star Formation from Converging Filaments

In W33A, two localized regions of star formation (MM1 and MM2) separated by 0.1 pc are surrounded by a common filamentary structure of ~ 1 pc in length (Sect. 2.4.2). The two velocity components of this filamentary structure are separated by ≈ 2.6 kms in line-of-sight velocity and intersect in projection right at the position of the star formation activity (Fig. 2.2). The velocity components are not a mere superposition in position–position–velocity space, but they merge into a structure with larger motions (Fig. 2.8, Sect. 2.4.2), resolved into a disk/outflow system by subarcsecond resolution observations (Figs. 2.5 and 2.6, Sect. 2.4.2). This suggests that star formation in W33A was triggered by the convergence of molecular filaments. Such a scenario has been suggested for the region W3 IRS 5 by [171]. More recently, the merging of filaments has also been claimed by [99] in the infrared dark cloud G35.39-0.33 and by [40] in the MSFR W75.

This mode of star formation is predicted by numerical simulations of star formation triggered by converging flows [15, 85, 208]. In those simulations, the formation of molecular clouds itself is a product of the convergence of streams of neutral gas. Later in the evolution of the molecular clouds, filaments of molecular gas can converge (merge) with each other, leading to the formation of cores and stars.

We present here a simple comparison with a region found in the numerical simulation recently reported by [209]. This simulation represents the formation of a giant molecular complex from the convergence of two streams of warm neutral gas, at the scale of tens of parsecs. Specifically, the simulation was performed using the smoothed particle hydrodynamics (SPH) code GADGET [194], including

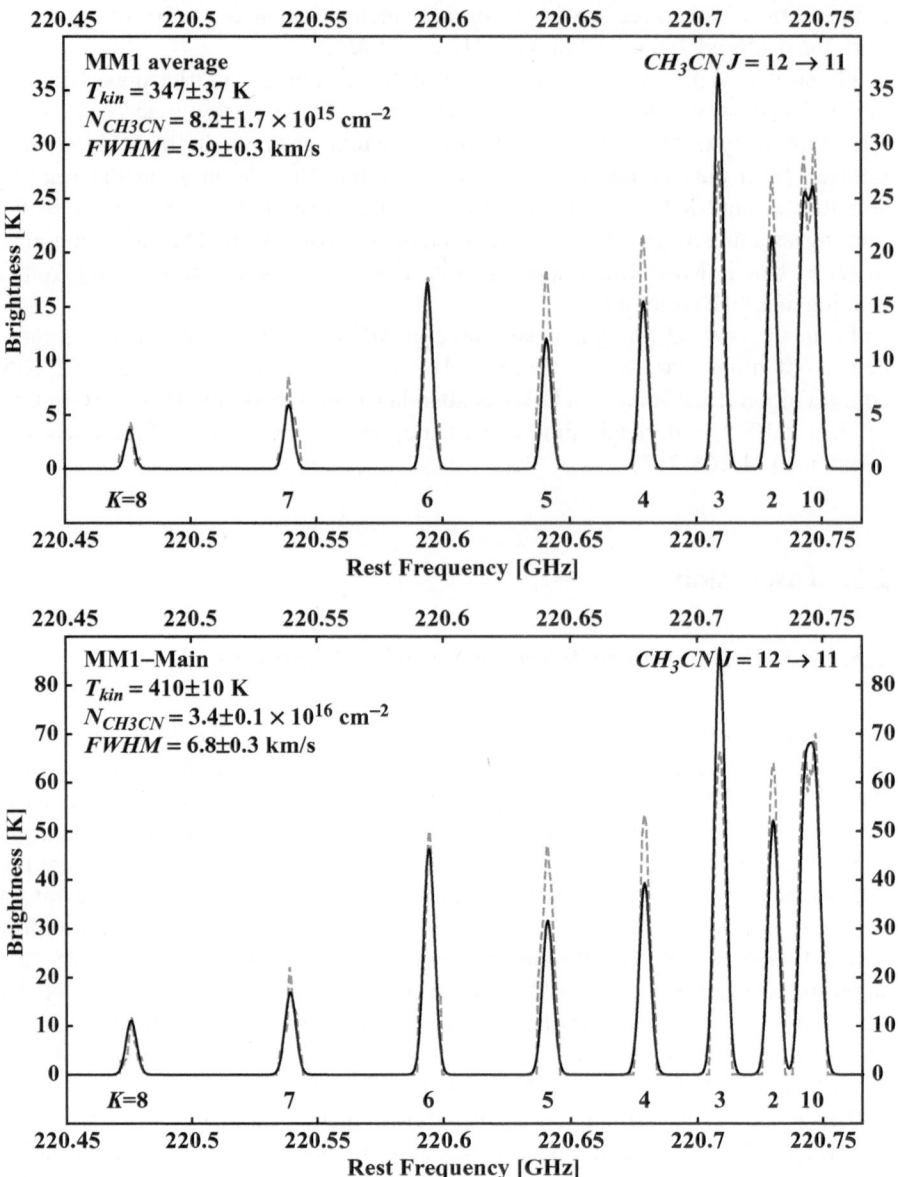

Fig. 2.8 Three-dimensional (position–position–velocity) rendering of the NH_3 (2,2) data. Every voxel in the data cube with intensity >20 mJy beam^{-1} (5σ) has been included. The vertical axis is color-coded according to V_{LSR}. It is seen that the two filamentary structures at different velocities do not merely superpose at the center position, but merge in position–position–velocity space, suggesting interaction of the filaments

sink-particle and radiative cooling prescriptions [97, 208]. The convergence of the warm diffuse-gas streams triggers a thermal instability in the gas, which causes it to undergo a transition to the cold atomic phase, forming a cloud. The latter soon becomes gravitationally unstable, begins contracting, and undergoes hierarchical fragmentation. During the contraction, the density of isolated clumps increases and they can reach physical conditions corresponding to those of molecular clumps. Finally, the global collapse reaches the center of mass of the cloud, at which point a region with physical conditions corresponding to those of MSFRs forms. The simulation box has 256 pc per side, and the converging flows have a length of 112 pc, and a radius of 32 pc. However, after the gravitational contraction, the clumps are only a few parsecs across. Since SPH is essentially a Lagrangian method, it allows sufficient resolution in these dense regions. We refer the reader to [208, 209] for details of the simulation. Here we focus on the region called "Cloud 1" in [209], albeit roughly 1.5 Myr later than the time examined in that paper.

Figure 2.9 shows two snapshots of column density separated by 0.133 Myr (the time interval between successive data dumps of the simulation). The column density is computed by integrating the density along the x-direction over the 10 pc path $123 \leq x \leq 133$ pc, which is centered at the midplane of the simulation, where the (sheet-like) cloud is located. In this region and epoch of the simulation, the two leftmost filaments in the *top panel* of Fig. 2.9 converge to form the boomerang-shaped filament seen in the *bottom panel*. Note that the simulation was not designed to simulate the observed filaments. The comparison is only intended to show that some of observed properties of W33A can arise naturally in the context of a simulation of the formation of a large molecular complex that contains a filament system.

The peaks of two filaments of gas are initially separated by 0.4 pc (*top frame* of Fig. 2.9) and then merge at a projected velocity of 3 ± 1.5 km s^{-1}, measured directly from the displacement observed between the two panels. The column density of the filaments is in the range $[10^3, 10^4]$ code units (Fig. 2.9), or $N_{H_2} = [0.5 \times 10^{24}, 0.5 \times 10^{25}]$ cm^{-2}. For an NH$_3$ abundance with respect to H$_2$ in the range $X(NH_3) = [10^{-8}, 10^{-7}]$ (e.g., [74]), the column density of the cold filaments in our observations is $N_{H_2} = [1.1 \times 10^{24}, 1.1 \times 10^{25}]$ cm^{-2}. Figure 2.10 shows the volume density (color scale) and y–z plane velocity (arrows) in a slice trough the filaments. It is seen that the filaments reach densities typical of MSFRs (peak $n \sim 10^5$ cm^{-3}) and that their velocity field presents fast jumps of a few km s^{-1} in the interaction zones, comparable to our observations. We conclude that some of the properties of the observed filaments such as sizes, column densities, and velocities agree within a factor of 2 with those from the simulation. This rough comparison illustrates that our interpretation of convergence between the observed filaments is feasible.

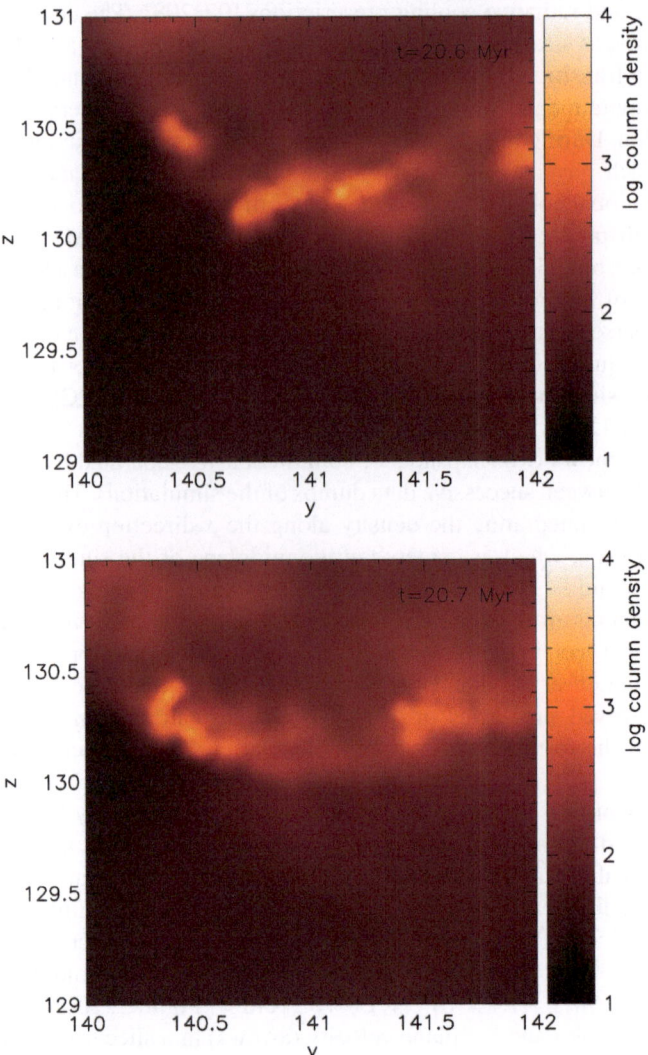

Fig. 2.9 Snapshots of molecular filaments merging with each other extracted from the simulation presented by Vázquez-Semadeni et al. [209]. The time interval between the first (*top*) and second (*bottom*) frames is 0.133 Myr. The units of the z- and y-axes are pc. The *color scale* shows the column density in code units, equivalent to 4.9×10^{20} H_2 particles per square cm

2.5.2   *Cores at Different Evolutionary Stages*

The star-forming cores in W33A appear to be at markedly different evolutionary stages (Sect. 2.4.2). The first piece of evidence for this is the clear difference in the richness of the molecular-line emission from MM1 to MM2 (see Fig. 2.4).

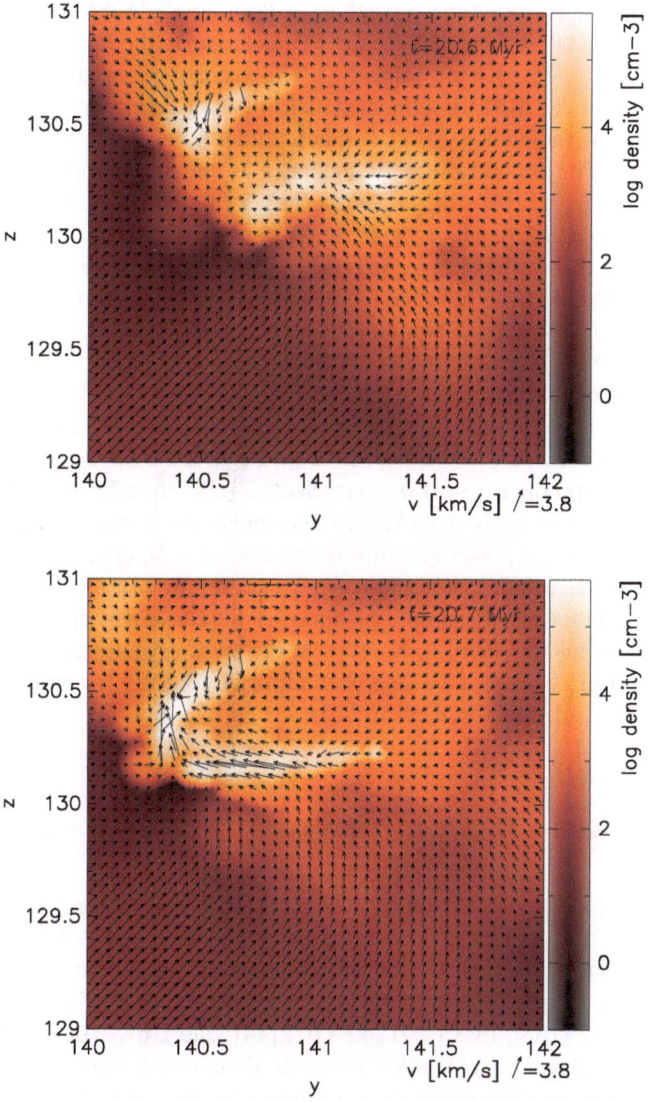

Fig. 2.10 Snapshots of molecular filaments merging with each other extracted from the simulation presented by Vázquez-Semadeni et al. [209]. The time interval between the first (*top*) and second (*bottom*) frames is 0.133 Myr. The units of the z- and y-axes are pc. The *color* scale shows the volume density (cm^{-3}) in a slice passing through the merging filaments. The *arrows* indicate the velocity of the gas in the y–z plane. In the second frame (*bottom*), the two merged filaments have velocities roughly opposing each other by a few km s^{-1}

MM1 has molecular emission typical of a "hot-core" [126], with a prominent CH_3CN $J = 12 - 11$ "K-forest" that can be detected up to the $K = 8$ line, with upper-level energy $E_u = 525.5$ K (see Table 2.3). The average gas temperature of MM1 is $\sim$347 K. In contrast, MM2 is almost devoid of "hot-core" emission, and is only detected in a few molecules. MM2 is much colder than MM1, with a temperature $T_{MM2} \approx 46$ K. The second piece of evidence is the mass content of the cores. MM1 has only $\sim[9,32]\,M_\odot$ of gas, while MM2 has $\sim$60$M_\odot$. This could naively be interpreted as MM2 having a much larger gas reservoir than MM1, but it should be kept in mind that both cores are part of a common parsec-scale structure. Clump infall at pc scales has been reported from single-dish [222] and interferometric [74] observations. Also, numerical simulations of parsec-sized clumps show that the star-forming cores that give birth to massive stars are continuously fed from gas in the environment at the clump scales (e.g., [32, 161, 186, 209]).

There are two possibilities that we briefly discuss here: (1) the prestellar cores that were the precursors to MM1 and MM2 appeared at the same time and then MM1 evolved faster to produce a $\sim$10$M_\odot$ star (Sect. 2.4.2), while MM2 only produced at most an intermediate mass star (MM2 is not prestellar, since it has some internal heating and appears to power an outflow, see Sect. 2.4.2), or (2) the prestellar core precursor to MM2 formed later and has yet to form at least one massive star. Our observations cannot tell these options apart. A measurement of the accretion rate in both cores would be helpful. Sources at different evolutionary stages within a single star-forming cluster have also been reported recently for AFGL 5142 [235], G28.34+0.06 [236], and AFGL 961 [216].

2.5.3 A Rotating Disk/Outflow System in MM1-Main

In the past decade, the question of whether massive stars form by disk-outflow mediated accretion similar to low-mass stars has been the subject of intensive research. The answer is positive: they definitely do. Some of the massive protostars that have been shown to harbor disk/outflow systems are G192.16-3.82 [182], Cepheus A HW2 [158], IRAS 20126+4104 [44], and IRAS 16547-4247 [68]. All these relatively clean disk examples, however, do not have stars more massive than 15–20$M_\odot$. More massive (O-type) stars have also been shown to form via disk-mediated accretion. However, the innermost part of the accretion flow is often (at least partially) ionized and is observed as a hypercompact (HC) H II region. Also, the gas in these more massive regions is warmed up to farther distances, and very massive rotating structures of size up to 0.1 pc are typically observed. Examples are G10.6-0.4 [108], G24.78+0.08 [22], W51e2 [117], and G20.08-0.14N [74], all of which have stellar masses above 20$M_\odot$. A possible exception to this scenario is W51 North, where [227, 228] claims to have found a protostar with $M_\star > 60M_\odot$ and without a "bright" (with flux above tens of mJy at wavelengths $\sim$1 cm) H II region. This apparent discrepancy is solved if multiple, lower-mass stars account for the mass in W51 North or if the H II region in this source is gravitationally

trapped as currently observed. Indeed, detailed simulations of the evolution of HC
H II regions within accretion flows show that their radio-continuum emission flickers
significantly in timescales from 10 to 10^4 year [161].

In this chapter we report on the existence of a rotating disk centered on MM1-
Main in W33A (Sect. 2.4.2). To our sensitivity, the radius of the disk is $R \lesssim$
4,000 AU. The warm-gas emission does not come only from the disk, but also
from an structure elongated perpendicular to it, coincident with the other mm peaks
(MM1-NW and MM1-SE) along the direction of the outflow (see Figs. 2.5 and 2.6).
We propose that the secondary mm peaks in MM1 are not of protostellar nature,
but regions where the emission is enhanced due to the interaction of the outflow
with the disk and its inner envelope. Indeed, the two brightest 7-mm detections of
[204] are counterparts of MM1-Main (Q1, $S_v \approx 1.7\,\text{mJy}$ at 7 mm) and MM1-SE
(Q2, $S_v \approx 0.6\,\text{mJy}$ at 7 mm).

The free–free emission from the 7 mm source Q1 should be a combination of
photoionization by the central protostar and shock-induced ionization of material
due to the jet observed by [55], likely dominated by the latter. The origin of the free–
free emission from Q2 could also be shocks, although deeper 7 mm observations
are needed to confirm this source. Q3 does not have a mm/submm counterpart and
may not be a distinct source. The radio continuum sources of W33A could then be
analogs of those in IRAS 16547-4247 [174, 175], which have fluxes a factor of a few
larger at 0.75 times the distance to W33A. Indeed, we find that the radio-continuum
emission from Q1 (MM1-Main) agrees with the correlation found for low-mass jets
between the radio-continuum luminosity of the jet and the momentum rate of the
associated molecular outflow: $\dot{P} = 10^{-2.5}(S_v d^2)^{1.1}$ [7], where the radio luminosity
has units of mJy kpc^2 and the momentum rate has units of $M_\odot\,\text{year}^{-1}\,\text{km s}^{-1}$. For
W33A with a radio flux $S_{3.6\text{cm}} = 0.79\,\text{mJy}$ [169], and a distance $d = 3.8\,\text{kpc}$ [96],
the expected momentum rate is $\dot{P} = 0.046\,M_\odot\,\text{year}^{-1}\,\text{km s}^{-1}$, while the observed
momentum rate (lower limit) is $\dot{P} = 0.040\,M_\odot\,\text{year}^{-1}\,\text{km s}^{-1}$ (obtained dividing
the momentum of the molecular outflow by its length, including the inclination
correction). This correlation was found to hold for three well-studied massive
protostars (IRAS 16547-4247, HH 80-81, and Cep A HW2) by [175]. In this study
we report that it also holds for W33A, which constitutes further evidence for a
common accretion mechanism between low- and high-mass protostars, at least to
the stage prior to the development of a brighter H II region.

Comparing the ratio of the radio luminosity to the IR luminosity $L(8\,\text{GHz})/L(\text{IR})$
with the recombination-line line width may be a useful criterion to distinguish
between a source ionized by shocks (jet or stellar wind) or by photoionization (what
usually is called an H II region). In Fig. 2.5 of the review by [90], it is seen that
UC H II regions have the largest $L(8\,\text{GHz})/L(\text{IR})$ and the smallest line width, jet
sources have the smallest $L(8\,\text{GHz})/L(\text{IR})$ and the largest line width, and HC H II
regions fall in between the previous two. For W33A, $\log(L(8\,\text{GHz})/L(\text{IR})) \approx 7.1$,
and the FWHM of the IR recombination lines is of several hundreds km s^{-1}, again
consistent with a jet source.

The star(s) at the center of MM1-Main (with $M_\star \sim 10\,M_\odot$) appears to dominate
the dynamics of the disk, but we cannot rule out the existence of additional, less

massive objects within it. Indeed, models of massive protostellar disks predict their fragmentation and the formation of a few lower mass companions within it [119, 122, 161].

2.6   Conclusions

In this chapter we presented for the first time resolved observations in both mm continuum and molecular-line emission for the massive star formation region W33A, characterized by a very high luminosity ($L \sim 10^5 L_\odot$) and very low radio-continuum emission (~ 1 mJy). Both of the previously known mm cores (MM1 and MM2) are resolved into multiple peaks, and appear to be at very different evolutionary stages, as indicated by their molecular spectra, masses, temperatures, and continuum spectral indices. The brightest core (MM1-Main at the center of MM1) is centered on a very faint free–free source and the gas dynamics up to a few thousand AU of it indicates the presence of a circumstellar disk rotating around a stellar mass of $M_\star \sim 10 M_\odot$. MM1-Main also drives a powerful, high-velocity molecular outflow perpendicular to the disk. MM2, the coldest and most massive core, is not detected in hot-core lines but appears to drive a more modest outflow. Both MM1 and MM2 are located at the intersection of parsec-scale filamentary structures with line-of-sight velocity offset by ≈ 2.6 km s^{-1}. Analysis of the position–position–velocity structure of these filaments and a comparison with recent numerical simulations suggests that star formation in W33A was triggered by the convergence of filaments of cold molecular gas.

Chapter 3
A MSFR with Young UC and HC HII Regions: G20.08N

3.1 Summary

Spectral line and continuum observations of the molecular and ionized gas in G20.08–0.14N explore the dynamics over a range of spatial scales in this massive star-forming region. Observations of NH_3 performed with the Very Large Array at $4''$ angular resolution show a large-scale ($\sim$0.5 pc radius) molecular accretion flow around and into a star cluster with three small H II regions. Higher resolution ($0.4''$) Submillimeter Array and VLA observations in hot core molecules (CH_3CN, OCS, and SO_2, NH_3) show that the two brightest and smallest H II regions are themselves surrounded by smaller scale ($\sim$0.05 pc radius) accretion flows. The rotation of the large-scale flow is aligned with the small scale, and the timescale for the contraction of the cloud is short enough, 0.1 Myr, for the large-scale flow to deliver significant mass to the smaller scales within the star formation timescale. The flow structure appears to be continuous and hierarchical from larger to smaller scales.

Millimeter recombination line (RL) observations at $0.4''$ angular resolution indicate rotation and outflow of the ionized gas within the brightest H II region (A). The broad recombination lines suggest supersonic flows. A continuum spectral energy distribution (SED) that rises from cm to mm wavelengths indicates a density gradient inside this H II region. These observations are consistent with photoevaporation of the inner part of the rotationally flattened accretion flow.

These results have been published in Galván-Madrid, Roberto, Keto, Eric R., Zhang, Qizhou, Kurtz, Stan, Rodríguez, Luis F., and Ho Paul T. P. "Formation of an O-Star Cluster by Hierarchical Accretion in G20.08–0.14N". The Astrophysical Journal, 706, 1036 (December 2009) [74].

3.2 Introduction

Massive star-forming regions (MSFRs) with O stars are usually identified by a group of hypercompact (HC) H II or ultracompact (UC) H II regions found together, deeply

R.J. Galván-Madrid, *On the Formation of the Most Massive Stars in the Galaxy*,
Springer Theses, DOI 10.1007/978-1-4614-3308-8_3,
© Springer Science+Business Media New York 2012

embedded in a dense molecular cloud [46, 77, 90]. That several H II regions are typically found within each star-forming region indicates that massive stars form together in small clusters. Furthermore, the infrared luminosity and radio continuum brightness of the individual H II regions suggest that some of them may themselves contain more than one massive star. Thus, the spatial structure of massive star-forming regions is clustered and hierarchical: the star-forming regions contain a number of separate HC and UC H II regions, each of which may in turn contain a few massive stars.

Low angular resolution, single-dish, molecular line surveys of MSFRs show evidence for large scale contraction of the embedding molecular clouds [116, 222]. Higher angular resolution observations of some of these regions identify velocity gradients consistent with rotation and inflow. In addition to the accretion flows seen on the large-scale ($\sim$0.3–1 pc) of the embedding molecular cloud (G10.6–0.4 [86, 103, 110]; G29.96–0.02 [155]), accretion flows are also seen on smaller ($\leq$0.1 pc) scales around individual HC and UC H II regions (G10.6–0.4 [112, 189]; W3(OH) [111, 113]; W51e2 [226, 230]; G28.20–0.05 [190]; G24.78+0.08 [20–22, 73]; G29.96–0.02 [30]).

It is unclear how the flows on different length scales are related. In the case of G10.6–0.4, the cluster-scale accretion flow can be traced down from the largest cloud scale to the small scale of the brightest H II region, but it is not known whether this holds for other objects. For example, in a survey of MSFRs, selected on the basis of IRAS colors and specifically excluding those with H II regions, multiple bipolar molecular outflows (implying the presence of accretion flows) are seen in random orientations [27–29]. The different orientations of these smaller-scale flows suggest separate, individual centers of collapse. This comparison raises the question whether a large-scale coherent flow is required for the formation of the most massive stars, O stars ($M_\star > 20 M_\odot$) capable of producing bright H II regions, whereas B stars require only smaller scale flows.

It is also unclear what happens in an accretion flow when the inflowing molecular gas reaches the boundary of an embedded H II region. Previous observations suggest that the H II regions in an MSFR that are surrounded by accretion flows, may be best understood as deriving from the continuous ionization of the accretion flow [104–106], rather than as a dynamically separate expanding bubble of ionized gas within the flow. Part of the ionized gas may continue to the central star or stars and part escapes off the rotationally flattened accretion flow as a photoevaporative outflow perpendicular to the plane of rotation [93, 100, 137, 225]. The outflow is accelerated to supersonic speeds by the density gradient maintained by the stellar gravity [106]. Because the extent of an ionized outflow is generally larger than the region of ionized inflow, in most cases the outflow should be detected more easily than the inflow. H II regions classified as "broad recombination line objects" (BRLO) [95, 180] show steep density gradients and supersonic flows [114], consistent with photoevaporation and acceleration. It is not known whether all BRLO are associated with accretion. If the accretion surrounding an O star cluster is continuous from the largest to the smallest scales, this must be the case.

There are only a handful of radio recombination line (RRL) observations that spatially resolve the ionized flow within an HC H II region. Velocity gradients consistent with outflow and rotation in the ionized gas have been previously reported for W3(OH) [113], W51e2 [107], and G28.20–0.05 [181]. Observations of the very massive and spatially large G10.6–0.4 H II region made at the VLA in the highest possible angular resolution are able to map the inflowing ionized gas [108].

In order to study the accretion dynamics over a range of scales in a MSFR, from the cluster scale down to the scale of individual HC H II regions and within the ionized gas, we set up a program of radio frequency molecular line, recombination line, and continuum observations at two telescopes and with several different angular resolutions. For the study presented in this chapter, we chose the massive star formation region G20.08–0.14 North (hereafter G20.08N), identified by three UC and HC H II regions detected in the cm continuum by Wood and Churchwell [221]. The total luminosity of the region is $L \sim 6.6 \times 10^5 L_\odot$ for a distance of 12.3 kpc.[1]

Previous observations suggest accretion in the G20.08N cluster. Molecular-line observations show dense gas embedding the H II regions [165, 202]. Molecular masers, generally associated with ongoing massive-star formation, have been detected in a number of studies (OH [88]; H_2O [91]; and CH_3OH [210]). Klaassen and Wilson [115, 116] observed large-scale inward motions consistent with an overall contraction of the embedding molecular cloud. Those authors also observed SiO line profiles suggestive of massive molecular outflows, further evidence for accretion and star formation. The recombination line spectra show broad lines [78, 180], presumably due to large, organized motions in the ionized gas. However, the previous observations do not have the angular resolution and the range of spatial scales needed to confirm the presence of accretion flows and study them in detail.

In this chapter we present several observations of G20.08N and discuss our findings. We confirm active accretion within the cluster. Furthermore, we find that the parsec-scale accretion flow fragments into smaller flows around the individual HC H II regions, and that the gas probably flows from the largest scale down to

---

[1] Both near and far kinematic distances have been reported for G20.08N. The near value given by Downes et al. [60] ($d \approx 4.1$ kpc) is the most commonly quoted in the previous literature. In contrast, Fish et al. [65] and Anderson and Bania [5] report that this region is at the far kinematic distance ($d \approx 12.3$ kpc). We will assume the far distance throughout the rest of this study. For reference, a scale of $0.5''$ corresponds to $\approx$6,000 AU (0.03 pc). The total luminosity of the region was estimated to be $L \sim 7.3 \times 10^4 L_\odot$ assuming the near kinematic distance [221]. Correcting for the location at the far distance, the luminosity is $L \sim 6.6 \times 10^5 L_\odot$.

the smallest scale. This continuous and hierarchical accretion may be necessary to supply enough mass to the small-scale flows to form O-type stars, in contrast to low- and intermediate-mass star-forming regions with stars no more massive than $\sim 20 M_\odot$, where isolated accretion flows around individual protostars may be sufficient.

3.3   Observations

3.3.1   SMA

We observed G20.08N on 2006 June 25 and July 6 with the Submillimeter Array[2] [89] in its very extended (VEX) configuration. Two sidebands covered the frequency ranges of 220.3–222.3 and 230.3–232.3 GHz with a spectral resolution of ≈ 0.5 km s^{-1}. The H30α recombination line ($\nu_0 = 231.9009$ GHz) was positioned in the upper sideband.[3] The observations sampled baseline lengths from ≈ 50 to $\lesssim 400$ kλ, sensitive to a range of spatial scales from $\approx 0.5''$ to $\approx 4.1''$.

The visibilities of each observation were separately calibrated using the SMA's data calibration program, MIR. Table 3.1 lists relevant information on the calibrators. We used quasars for the absolute amplitude scale as well as the time-dependent phase corrections and frequency-dependent bandpass corrections. Inspection of

Table 3.1 Observational parameters

Epoch	Array	Phase center[a] α(J2000)	δ(J2000)	Bandpass calibrator	Phase calibrator	Flux calibrator
2003 Apr 28	VLA-D	18 28 10.384	−11 28 48.65	3C454.3	1,833 − 210	0137 + 331
2003 May 13	VLA-D	18 28 10.384	−11 28 48.65	3C454.3	1,851 + 005	1,331 + 305
2003 Oct 09	VLA-BnA	18 28 10.384	−11 28 48.65	3C454.3	1,851 + 005	1,331 + 305
2003 Oct 10	VLA-BnA	18 28 10.384	−11 28 48.65	3C454.3	1,851 + 005	1,331 + 305
2006 Jun 25	SMA-VEX	18 28 10.38	−11 28 48.60	3C273	1,830 + 063	1,830 + 063
2006 Jun 25	SMA-VEX	18 28 10.76	−11 29 27.60	3C273	1,830 + 063	1,830 + 063
2006 Jul 06	SMA-VEX	18 28 10.38	−11 28 48.60	3C454.3	1,751 + 096	1,751 + 096
2006 Jul 06	SMA-VEX	18 28 10.76	−11 29 27.60	3C454.3	1,751 + 096	1,751 + 096
2007 Oct 26	VLA-B	18 28 10.400	−11 28 49.00	3C454.3	1,743 − 038	1,331 + 305
2007 Oct 27	VLA-B	18 28 10.400	−11 28 49.00	3C454.3	1,743 − 038	1,331 + 305

[a]Units of R.A. are hours, minutes, and seconds. Units of decl. are degrees, arcminutes, and arcseconds

---

[2]The Submillimeter Array is a joint project between the Smithsonian Astrophysical Observatory and the Academia Sinica Institute of Astronomy and Astrophysics and is funded by the Smithsonian Institution and Academia Sinica.

[3]In "chunk" 20 of the SMA correlator setup.

Table 3.2 Lines[a]

Species	Transition	ν_0 (GHz)	Array	HPBW (arcsec $\times$ arcsec; deg)
H	66α	22.364178	VLA-B	0.47×0.34; 3
NH_3	(2,2)	23.722633	VLA-D	4.68×2.93; 236
NH_3	(3,3)	23.870129	VLA-D	4.50×3.32; 8
NH_3	(2,2)	23.722633	VLA-BnA	0.37×0.28; 343
NH_3	(3,3)	23.870129	VLA-BnA	0.71×0.34; 0
^{13}CO	2–1	220.398681	SMA-VEX	0.55×0.41; 32
CH_3CN	12(4)–11(4)	220.679297	SMA-VEX	0.55×0.41; 32
CH_3CN	12(3)–11(3)	220.709024	SMA-VEX	0.55×0.41; 32
CH_3CN	12(2)–11(2)	220.730266	SMA-VEX	0.55×0.41; 32
SO_2	11(1,11)–10(0,10)	221.965200	SMA-VEX	0.54×0.41; 32
CO	2–1	230.538000	SMA-VEX	0.53×0.39; 37
OCS	19–18	231.060991	SMA-VEX	0.53×0.39; 37
H	30α	231.9009	SMA-VEX	0.53×0.39; 37
CH_3CN	12(7)–11(7)	220.539340	SMA-VEX	0.55×0.41; 32
HNCO	10(1,9)–9(1,8)	220.584762	SMA-VEX	0.54×0.41; 32
CH_3CN	12(6)–11(6)	220.594438	SMA-VEX	0.55×0.41; 32
CH_3CN	12(5)–11(5)	220.641096	SMA-VEX	0.55×0.41; 32
CH_3CN	12(1)–11(1)	220.743015	SMA-VEX	0.55×0.41; 32
CH_3CN	12(0)–11(0)	220.747265	SMA-VEX	0.55×0.41; 32
CH_2CHCN	24(0,24)–23(0,23)	221.76598	SMA-VEX	0.55×0.41; 32
^{13}CS	5–4	231.220768	SMA-VEX	0.52×0.39; 37

[a]Lines detected at S/N > 6. The *top* part of the table lists the lines with S/N > 10 (except the H66α line) and clearly isolated in frequency. The *bottom* part of the table lists the lines detected at S/N < 10 or blended. Some spectral features (see Fig. 3.1) that were not properly identified due to low S/N ($\approx$5) and blending are not listed

the quasar fluxes and comparison with their historical flux densities in the SMA database[4] suggest that flux calibration is accurate to better than 20%. The calibrated data were exported to MIRIAD for further processing and imaging.

There were enough line-free channels in the 2-GHz passband to subtract the continuum in the (u,v) domain. The line-free continuum was self-calibrated in phase, and the gain solutions were applied to the spectral line data. We list all the identified lines in Table 3.2. Figure 3.1 shows the continuum-free spectra across both sidebands at the position of the 1.3-mm peak.

To improve the sensitivity, the data were smoothed to a spectral resolution of 2 km s^{-1}. The rms noise in our natural-weighted maps, made from the combined observations of both days, is $\sim$2 mJy beam^{-1} for the single-sideband continuum and $\sim$30 mJy beam^{-1} per channel (2 km s^{-1} wide) for the line data.

---

[4]http://sma1.sma.hawaii.edu/callist/callist.html

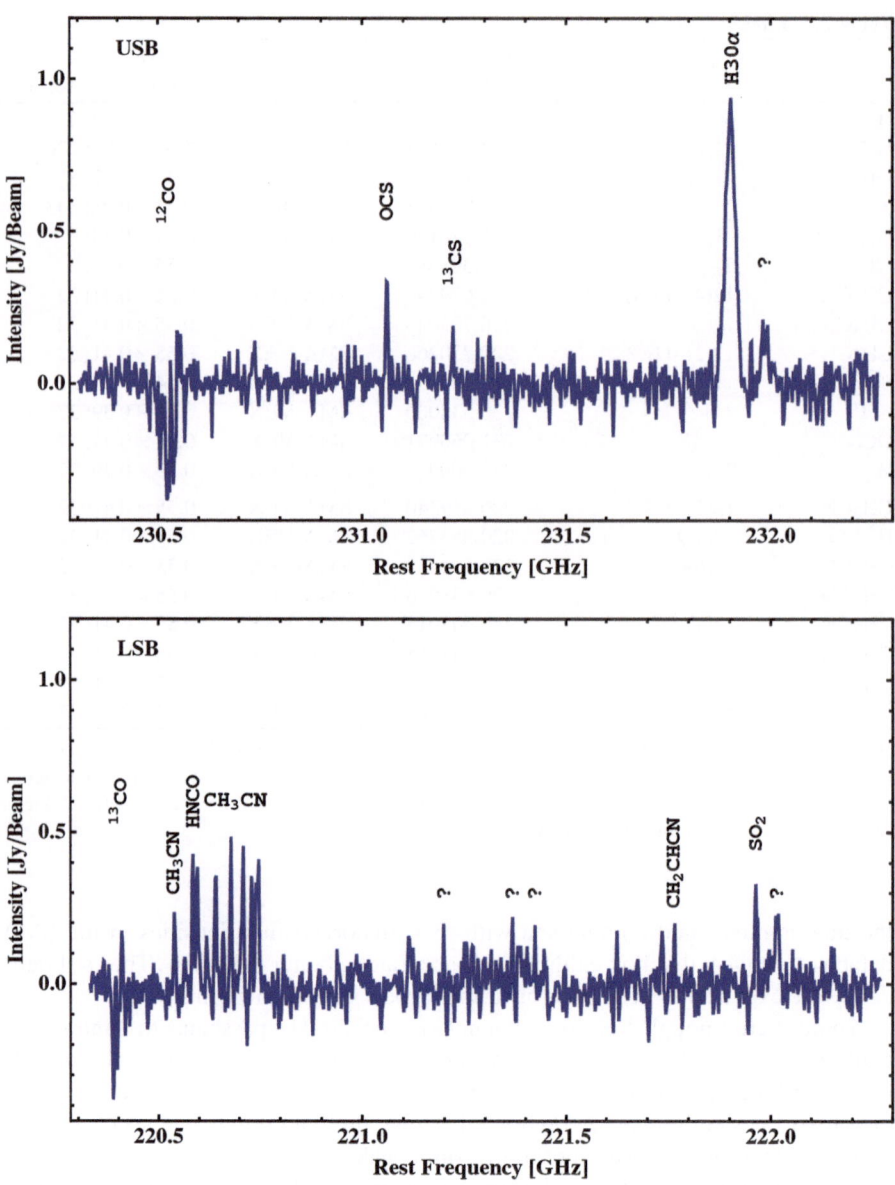

Fig. 3.1 Wide-band, continuum-free spectra from the SMA-VEX data at the position of the 1.3-mm continuum peak (see Table 3.2 for details). The channel spacing in this plot is 3 km s^{-1}. The question mark (*?*) in the upper sideband (*top frame*) might be a superposition of lines of CH$_3$OCH$_3$ and CH$_3$CH$_2$CN. The *?* symbols at the center of the lower sideband (*bottom frame*) could be from vibrationally excited CH$_3$CN. The *?* symbol in the upper sideband close to the SO$_2$ might be from CH$_3$CCH

3.3.2 VLA

Spectral line observations of NH_3 $(J, K) = (2, 2)$ and $(3,3)$ at two different angular resolutions were obtained with the Very Large Array[5] in the D and BnA configurations (projects AS749, AS771, and AS785). Partial results from these observations were presented in Sollins [187]. All the observations, except the VLA-D $(3,3)$, were done with a bandwidth of 3.125 MHz ($\approx$39 km s^{-1}) divided in 64 spectral channels, each 0.6 km s^{-1} wide. The VLA-D $(3,3)$ observation was done with the same bandwidth divided in 128 spectral channels, each 0.3 km s^{-1} wide. The bandwidth covers the main hyperfine line and one line from the innermost satellite pair.[6] The NH_3 data are presented at a spectral resolution of 0.6 km s^{-1}. The noise per channel in the final images was in the range of 1.0–1.5 mJy beam^{-1}.

In addition to the molecular line observations, we observed the ionized gas in the H66α recombination line in the VLA B configuration. The correlator was set up to cover a bandwidth of 12.5 MHz ($\approx$166 km s^{-1}) divided in 64 channels of 2.6 km s^{-1} each. The rms noise per channel in the final image was $\approx$1 mJy beam^{-1}.

All three VLA data sets were calibrated using standard procedures in the AIPS software. Tables 3.1 and 3.2 summarize the relevant observational parameters. The continuum was constructed in the (u, v) domain from line-free channels and was then self-calibrated. The gain solutions from self-calibration were applied to the line data.

3.4 Results and Discussion

3.4.1 The Continuum Emission

3.4.1.1 Morphology

Figure 3.2 shows the 1.3-cm continuum (contours) obtained from the VLA-BnA observations overlaid with the 1.3-mm continuum from the SMA-VEX data (color scale). At 1.3 cm we resolve the G20.08N system into the three components reported by [221]. H II region A is the brightest, westernmost peak. H II region B is the slightly broader peak $\approx$0.7$''$ to the SE of A. H II region C is the more extended UC H II further to the SE. Its brightest, eastern rim is detected at the $\sim$10 mJy beam^{-1} level in our 1.3-mm observations.

The continuum of H II region A is unresolved at 1.3 cm; at 1.3 mm it shows a core-halo morphology. The 1.3-mm core is unresolved (Gaussian fits yield a deconvolved size at half power FWHM $\lesssim$ 0.4$''$). The low-intensity halo has a

---

[5]The National Radio Astronomy Observatory is operated by Associated Universities, Inc., under cooperative agreement with the National Science Foundation.

[6]The NH_3 molecule is symmetric top with inversion, see Ho and Townes [87] for details.

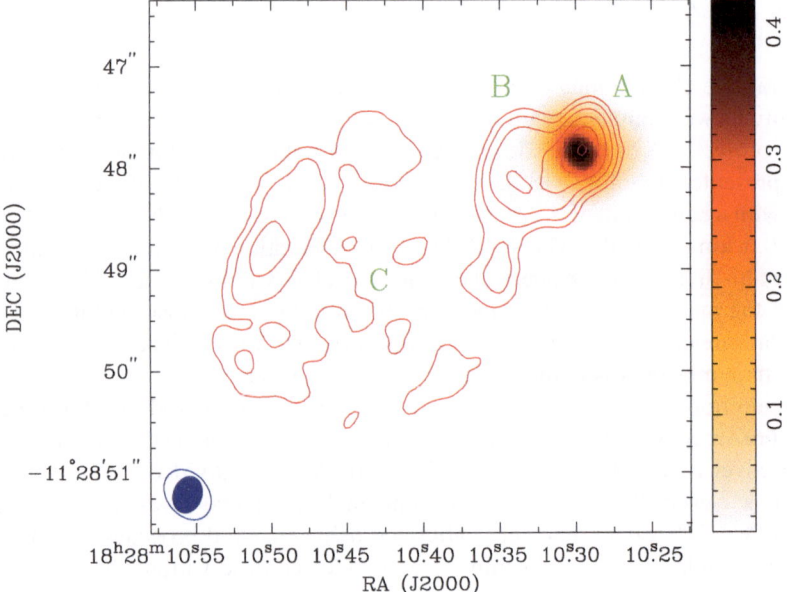

Fig. 3.2 VLA-BnA 1.3-cm continuum (*red contours*) overlaid on the SMA-VEX 1.3-mm continuum (*color scale*) toward the G20.08N complex. From west to east: H II region A is the compact, strongest peak at both wavelengths. H II region B is the less bright H II region at $\lesssim 1''$ to the SE of A. H II region C is the more extended emission to the SE of B. The color scale goes linearly from 8 to 430 mJy beam^{-1} (the rms noise in the mm image is 2 mJy beam^{-1}). Contours are placed at $-5, 5, 10, 20, 40, 80, 150 \times 1$ mJy beam^{-1}, the noise of the cm image. The SMA-VEX beam (*empty ellipse*) encircles the VLA-BnA beam (*filled ellipse*) at the *bottom-left* of the image

diameter of $\approx 1''$ (see Fig. 3.2). The H30α emission is confined to the unresolved core (Sect. 3.4.4) and the warm molecular gas (Sect. 3.4.3) coincides with the extended continuum halo. This indicates that the unresolved H II region A is surrounded by a dust cocoon. H II region B is barely resolved in the 1.3-cm VLA BnA map (deconvolved FWHM $\approx 0.6''$). The peak position of H II region A is identical at 1.3 cm and 1.3 mm: α(J2000) = 18^h 28^m 10^{s}30, δ(J2000) = $-11° 28' 47''.8$, within the positional uncertainty of the reference quasars (in the range $0.01''$–$0.1''$).

3.4.1.2 SED and the Nature of the Millimeter Emission

Previous observations of the recombination lines at 2 and 6 cm [180] put the H II regions in G20.08N in the class known as "broad recombination line objects" (although those observations did not have sufficient angular resolution to separate H II regions A, B, and C). The large widths in cm-wavelength recombination lines are due to pressure broadening at high gas densities ($>10^5$ cm^{-3}) as well as unresolved supersonic motions [114]. These H II regions also have continuum spectral energy distributions (SEDs) that increase with frequency through the mm wavelengths, evidence for a steep density gradient in the ionized gas [13,14,66,105,114]. Our new VLA and SMA observations, at 1.3 cm and 1.3 mm respectively, extend the SED to millimeter wavelengths. We find that the flux density of H II region A continues to rise from cm to mm wavelengths (Fig. 3.2), and we analyze this characteristic in detail below. The flux densities of H II region B at 1.3 mm ($\approx$93 mJy) and 1.3 cm ($\approx$202 mJy) imply a spectral index (α, where $S_\nu \propto \nu^\alpha$) of $\alpha \sim -0.3$, roughly consistent with the expected index of -0.1 of optically thin gas. H II region C is more extended and most of the 1.3-mm flux is resolved out.

At wavelengths shorter than 1 mm, thermal dust emission contributes significantly to the continuum. We estimate the relative contributions of dust and free–free emission at 1.3 mm from our recombination line observations (described in Sect. 3.4.4) and the theoretically expected line-to-continuum ratio. In the optically thin limit (a good approximation at 1.3 mm) the free–free line-to-continuum flux ratio S_L/S_C is given by the ratio of the opacities κ_L/κ_C [83]:

$$\left[\frac{\kappa_L}{\text{cm}^{-1}}\right] = \frac{\pi h^3 e^2}{(2\pi m_e k)^{3/2} m_e c} n_1^2 f_{n_1,n_2} \phi_\nu \frac{n_e n_i}{T^{3/2}} \times \exp\left(\frac{E_{n_1}}{kT}\right)\left(1 - e^{-h\nu/kT}\right), \quad (3.1)$$

$$\left[\frac{\kappa_C}{\text{cm}^{-1}}\right] = 9.77 \times 10^{-3} \frac{n_e n_i}{\nu^2 T^{3/2}}\left[17.72 + \ln\frac{T^{3/2}}{\nu}\right], \quad (3.2)$$

where all the units are in cgs, the physical constants have their usual meanings, $n_1 = 30$ for H 30α, $f_{n_1,n_2} \approx 0.1907 n_1(1 + 1.5/n_1)$ for α lines, $n_e = n_i$ for hydrogen, $T \approx 8{,}000$–$10{,}000$ K, and ϕ_ν is the normalized line profile. The main source of uncertainty in equations (3.1) and (3.2) is the temperature of the ionized gas. Assuming that the line profile is Gaussian and correcting for 8% helium in the gas, the expected ratio at the line center is $S_{L,0}/S_C \approx 3.0$ for $T = 8{,}000$ K, or $S_{L,0}/S_C \approx 2.3$ for $T = 10{,}000$ K. The observed ratio is $S_{L,0}/S_C \approx 1.8$. Therefore, assuming that the RRLs are in LTE, the free–free contribution to the 1.3-mm flux of H II region A is $\approx$60–80% for the assumed temperature range.

Figure 3.3 shows a model SED for H II region A in which 70% (355 mJy) of the 1.3-mm flux is produced by free–free from an H II region with a density gradient and 30% (142 mJy) by warm dust. Assuming radiative equilibrium, we set the dust temperature T_d to 230 K, the average temperature of the dense gas at the same scales (Sect. 3.4.3). The modeling procedure is described in Keto [105] and Keto et al. [114]. Table 3.3 summarizes the model. The total gas mass inferred from the dust emission is too large ($M \sim 35$–$95\,M_\odot$) for the H II region alone, so most of the dust must be in the cocoon around the H II region. The calculated mass range takes into account uncertainties in the dust emissivity, but not in the temperature.

The density gradient derived for the ionized gas in H II region A is $n_e \propto r^{-\gamma}$, with $\gamma = 1.3$. Equilibrium between recombination and ionization in this model H II

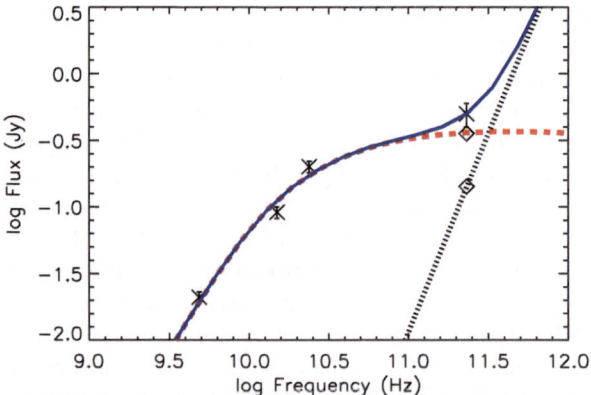

Fig. 3.3 Radio-to-millimeter SED of G20.08N A. The 6- and 2-cm measurements were obtained from Wood and Churchwell [221]. The 1.3- and 1.3-mm points were obtained from Gaussian fits to our VLA-BnA and SMA-VEX data, respectively. *Crosses* are the data points. The error bars correspond to the 10% and 20% uncertainty expected in the VLA and SMA flux measurements, respectively. The *red dashed line* shows the flux of an H II region with a density gradient. The *black dotted line* is the flux from the warm dust component. The *solid blue line* is the sum of the two components. The relative contributions of free–free (70%) and dust (30%) to the 1.3-mm flux were estimated from the observed H30α line-to-continuum ratio, and are marked with *diamonds*

Table 3.3 Model for the H II region G20.08N A

Parameter[a]	Value
HII radius (AU)	2,530
Electron density[b] (10^5 cm^{-3})	1.4
Exponent[c]	1.3
Gas mass[d] ($M_\odot$)	35–95
Spectral type	O7.5
Stellar mass ($M_\odot$)	34
HII mass ($M_\odot$)	0.05

[a]The independent parameters are the first four rows
[b]Electron density at the HII radius
[c]Exponent γ of the power-law density gradient in the ionized gas, where $n \propto r^{-\gamma}$
[d]Mass of molecular gas obtained from the dust emission. Assuming dust temperature $T_d = 230$ K and a gas-to-dust ratio of 100. The range is caused by the dust emissivity coefficient β used (see Keto et al. [114]), from $\beta = 1$ to $\beta = 1.5$

region requires an ionizing flux equivalent to an O7.5 star (using the computations of Vacca [203]), although this ionizing flux could be made up of several stars of slightly later spectral type. The model for the SED assumes spherical geometry and a static H II region with no inflow of neutral gas into the H II region. In contrast, H II region A is embedded in a rotationally flattened accretion flow (see Sect. 3.4.3), so the determination of the stellar spectral type is only approximate.

3.4.2 The Large-Scale Molecular Cloud

The large-scale molecular cloud is detected by the NH_3 VLA-D observations. From these data we find the presence of a parsec-scale accretion flow surrounding the cluster of H II regions. In the following sections we first estimate the systemic velocity V_{sys} of the cloud with respect to the local standard of rest (LSR), and then analyze the line velocities to determine rotation and infall following the procedure used in measuring accretion flow velocities by previous studies (e.g., [86, 110, 226, 230]). We also derive the physical properties of the cloud: temperature, mass, ammonia abundance, and density.

3.4.2.1 Dynamics

Figure 3.4 shows the channel maps of the $(J, K) = (3, 3)$ main hyperfine line. The most notable feature is a velocity gradient along the major axis of the cloud, consistent with rotation. The redshifted emission is toward the NE, while the blueshifted gas is toward the SW. Also, there is strong absorption against the free–free background of the H II regions at the center of the cloud, which are unresolved at $\sim 4''$ resolution. The separation between the emission peaks on either side of the absorption is $\approx 10''$, or 0.6 pc. The symmetry of the channel maps suggests a systemic velocity of about 42 km s^{-1}. A Gaussian fit to the emission spectrum integrated in a box covering the entire cloud, and clipping out the redshifted absorption, gives a centroid velocity of 42.3 ± 0.4 km s^{-1}. We adopt a systemic velocity for the parsec-scale cloud of $V_{sys} = 42$ km s^{-1}, in agreement with that reported by Plume et al. [165] (41.9 km s^{-1}) based on observations of CS $J = 7 - 6$.

Figure 3.5 shows position-velocity (PV) diagrams across cuts at PA $= 45°$ (SW–NE, major axis) and PA $= 135°$ (NW–SE, minor axis). The rotation is seen in the SW–NE cut as a shift in the velocity of the emission contours from one side of the absorption to the other. The velocity offset with respect to V_{sys} seen in the emission contours increases inward, suggesting that the gas rotates faster with decreasing radius.

Under the assumption that the velocity gradient seen in emission along the major axis is dominated by rotation, the redshifted absorption in the PV diagrams (Fig. 3.5) is also evidence for inward flow toward the central H II regions, as there is an excess of redshifted absorption. This is more clearly seen in Fig. 3.6, which shows the spectra toward the absorption center. The NH_3 (2,2) main hyperfine absorption

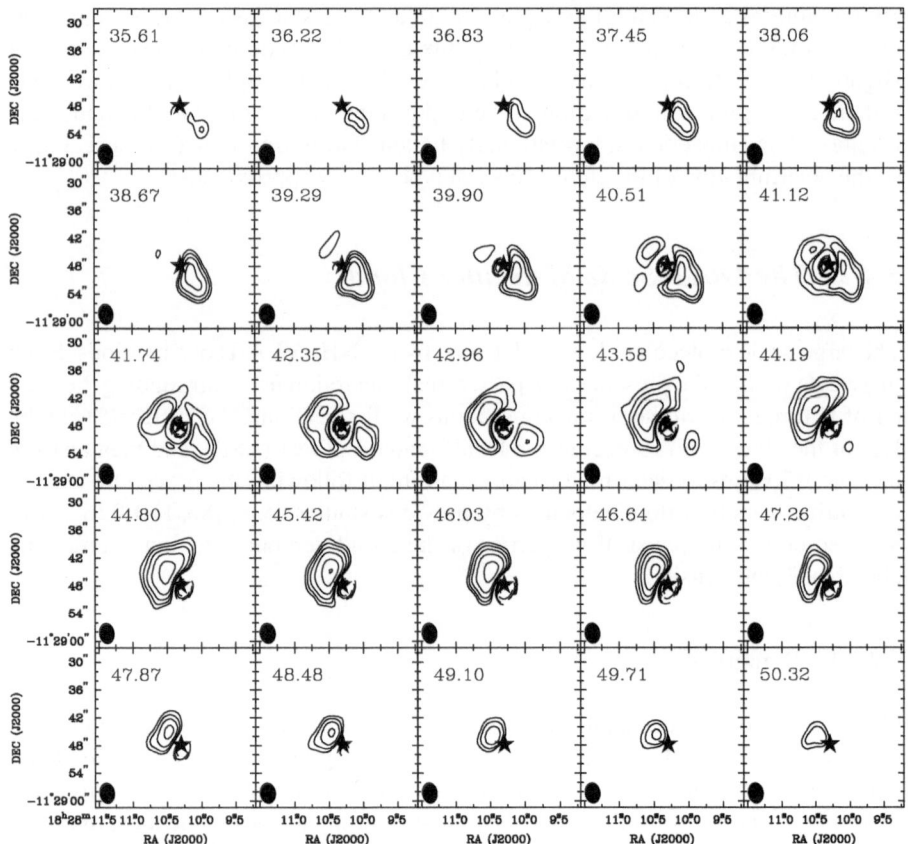

Fig. 3.4 Channel maps of the VLA-D NH$_3$ (3,3) observations. Emission is in *solid contours* and absorption in *dashed contours*. The star covers the H II regions shown in Fig. 3.1. Contour levels are at $-35, -25, -15, -10, -7, -5, 5, 7, 10, 15, 25, 35 \times 2$ mJy beam^{-1}. A clear velocity gradient in emission is seen from one side of the absorption to the other. The LSR systemic velocity of the molecular gas is $V_{sys} = 42.0$ km s^{-1}. The original maps at 0.3 km s^{-1} spectral resolution were smoothed to 0.6 km s^{-1} for clarity

peak is redshifted by 2.3 km s^{-1} with respect to the systemic velocity. An infall velocity $V_{inf} \approx 2$ km s^{-1} is also seen in the NH$_3$ (3,3) line, although the spectrum is contaminated by an NH$_3$ (3,3) maser (see description in Fig. 3.6). The maser is confirmed by our high angular resolution NH$_3$ observations (Sect. 3.4.6).

If the rotation were seen edge-on, there would not be a velocity gradient across the H II region along the minor axis (NW–SE). At an oblique viewing angle a velocity gradient along the minor axis is created by the inflow. The velocity gradient along both the minor and major axes (Fig. 3.5) implies that the rotationally flattened flow is tipped with respect to the line of sight.

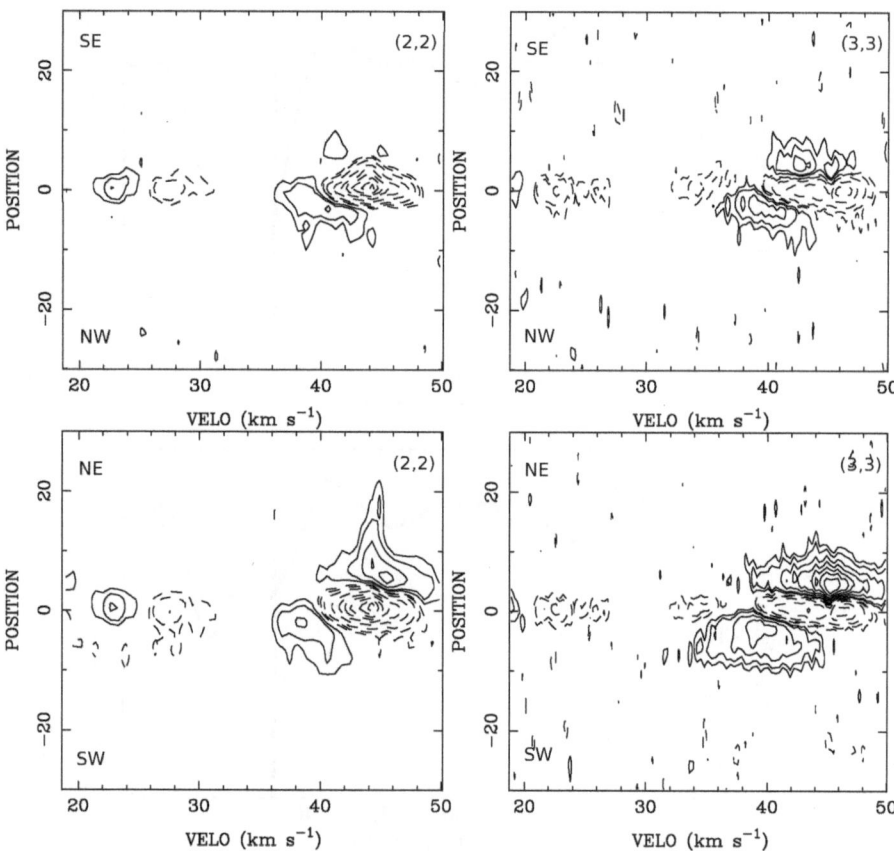

Fig. 3.5 Position-velocity diagrams of NH$_3$ (2,2) and (3,3) from the VLA-D data. The cuts were done at PA $= 45°$ (*bottom row*) and PA $= 135°$ (*top row*). *Dashed contours* are absorption, *solid contours* are emission. Contouring is at $-168, -144, -120, -96, -72, -48, -36, -24, -12, -4, 4, 8, 12, 16, 24, 32, 40, 48, 56, 64 \times 1$ mJy beam^{-1}. Only one inner satellite line is covered in the velocity range. In the SW–NE cuts (across the major axis of the cloud) the difference in the velocity of the emission with respect to $V_{sys} = 42$ km s^{-1} increases closer to the position center. This can be interpreted as spin up with decreasing distance from the center. However, the same trend is present in the NW–SE cuts (along the minor axis of the cloud), although only in the stronger, blueshifted side of the emission. This suggests that besides rotation, radial motions in the frame of the central stars are also present

3.4.2.2 Cloud Parameters

The optical depth of the gas can be determined from the brightness ratios of the hyperfine lines. From their optical depth ratio the rotational temperature between the (2,2) and (3,3) transitions can be determined (see Keto et al. [110] and Mangum et al. [141] for details of the procedure). This temperature can be considered as

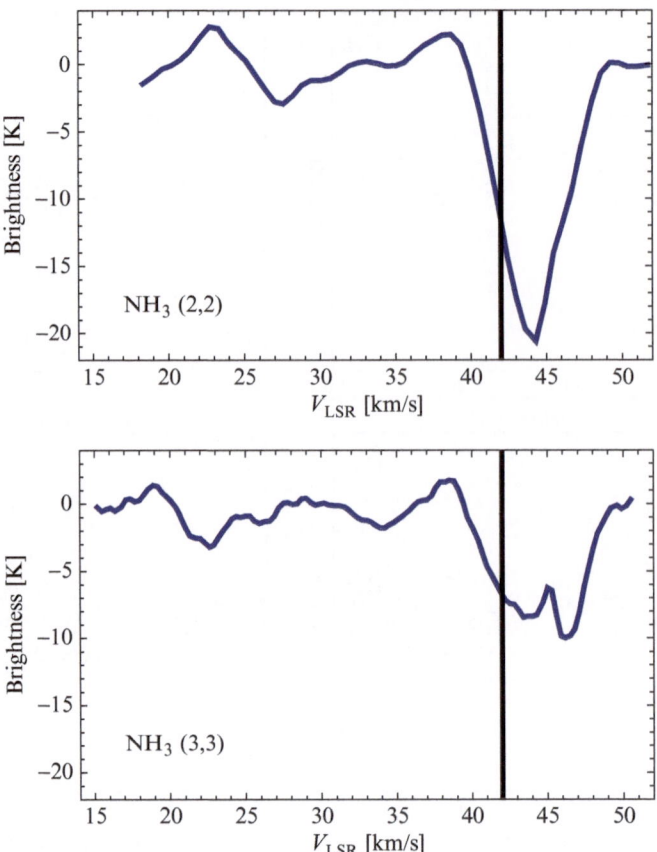

Fig. 3.6 NH$_3$ spectra from the VLA-D observations toward the center of the absorption in G20.08N. The *vertical line* marks the systemic velocity ($V_{sys} = 42$ km s^{-1}). The small peak near the middle of the (3,3) absorption is due to maser emission (confirmed in the high-resolution, VLA-BnA data, see Sect. 3.4.6). The absorption peaks in the main (2,2) and (3,3) lines are redshifted with respect to V_{sys}, indicating the presence of inflow in the kinematics of the parsec-scale molecular cloud. The other weak absorption component seen in the (2,2) spectrum is one of the inner satellites. The absorption component at $\approx$22 km s^{-1} in the (3,3) spectrum is also an inner satellite. The weaker absorption at 30–37 km s^{-1} in the (3,3) spectrum may arise from an outflow. The blueshifted absorption in the main lines closest to V_{sys} might arise from blending with the rotation seen in emission

a first-order approximation to the kinetic temperature (T_k) of the molecular gas. Danby et al. [54] suggest that an improved estimate of the kinetic temperature (T_k) is obtained by accounting for the populations in the upper states of the K-ladders rather than assuming that all states above the lowest are negligibly populated. Including this correction, we obtain an average kinetic temperature $T_k \sim 22$ K for the parsec-scale cloud. The high-resolution data discussed in Sect. 3.4.3 show that higher gas temperatures are found closer to the H II regions.

The mean column density of NH_3 is $N_{NH_3} \sim 2.3 \times 10^{16}\,cm^{-2}$. We can determine the abundance of ammonia $X(NH_3)$ by comparing the NH_3 and H_2 column densities. The mass of the cloud can be estimated from the observed velocity dispersion ($\sigma_V \sim 3.5\,km\,s^{-1}$), the radius of the cloud ($R \sim 100,000\,AU$), and the virial theorem: $M_{vir} = (5/3G)R\sigma_V^2 \sim 2,300\,M_\odot$. Similarly, from the observed velocity gradient, the rotation velocity is $V_{rot} \sim 4\,km\,s^{-1}$ at a radius $R \sim 100,000\,AU$. Equating the centripetal and gravitational forces, the gas mass inside R is $M = (RV_{rot}^2)/G \sim 1,800\,M_\odot$. Therefore, we estimate the average column density of H_2 to be $N_{H_2} \sim 5 \times 10^{22}\,cm^{-2}$ and $X(NH_3) \sim 5 \times 10^{-7}$, within a factor of 3 of the value estimated for G10.6–0.4 [103]. The mean H_2 density in the large-scale cloud is $n_{H_2} \sim 10^4\,cm^{-3}$.

3.4.3 Molecular Gas in the Inner 0.1 pc

Our SMA Very Extended (VEX) and VLA-BnA observations provide a rich view of the molecular environment close to the H II regions. The lines detected with the SMA that have peak intensities >300 mJy beam^{-1} and are not blended with any other line are listed in the top part of Table 3.2. We follow a similar outline as for the large-scale cloud.

3.4.3.1 Dynamics

Figures 3.7–3.10 show the channel maps of OCS $J = 19 - 18$, $SO_2\ J(K_a, K_b) = 11(1,11) - 10(0,10)$, $CH_3CN\ J(K) = 12(3) - 11(3)$, and $CH_3CN\ J(K) = 12(4) - 11(4)$. These maps show that the line emission from these hot-core molecules is considerably brighter around H II region A. All the molecules show a velocity gradient across this source, from the southwest to northeast, in a similar orientation as the larger, cluster-scale flow. We do not detect molecular emission around H II region B. At this high angular resolution we are not sensitive to brightness temperatures of less than $\sim 10\,K$ (3σ) for the mm lines.

Figure 3.11 shows the velocity-integrated intensity (moment 0) and the intensity-weighted mean velocity (moment 1) maps of the four lines previously shown in Figs. 3.7–3.10. The integrated emission of $CH_3CN\ 12(4) - 11(4)$ is brightest in front of the H II region, while for the other molecules the brightness peak is slightly offset from the continuum. These differences reflect the relative brightness of each molecule with respect to the continuum emission of H II region A. The velocity gradient across H II region A is also seen. Figure 3.12 shows the velocity-integrated (moment 0) and velocity dispersion (moment 2) maps for the same lines as Fig. 3.11. The line widths increase toward the continuum peak, indicating that unresolved motions increase closer to the H II region.

Figure 3.13 shows the position-velocity (PV) diagrams for the lines of Fig. 3.10 in cuts at position angles $PA = 45°$ and $PA = 135°$ across the continuum peak of HII

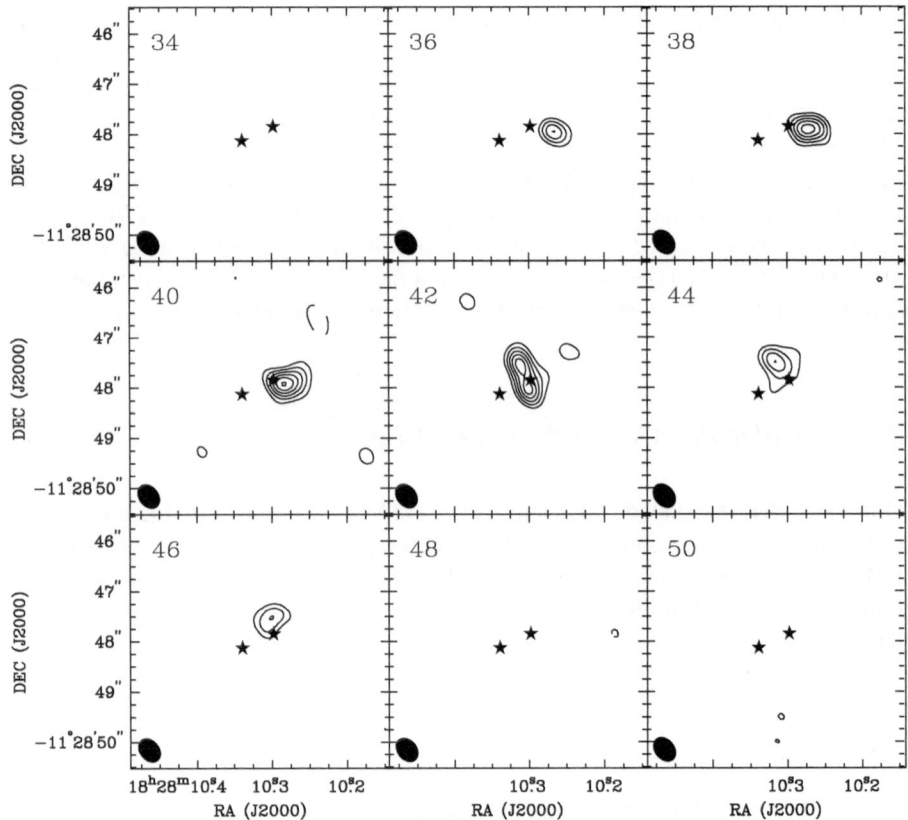

Fig. 3.7 Channel maps of OCS $J = 19 - 18$ from the SMA-VEX observations. Contours are $-4, 4, 6, 8, 10, 12, 14 \times 30$ mJy beam^{-1} (negative in *dashed*, and positive in *solid*). The peak intensity is 423 mJy beam^{-1}. The numbers in the *upper left corner* indicate the central LSR velocity of the channel. The *two stars mark* the positions of H II regions A (west) and B (east)

region A. The cuts at $45°$ show the velocity gradient also seen in the channel maps of OCS and both of the CH$_3$CN lines. The PV diagram of CH$_3$CN $12(4) - 11(4)$ has a feature suggestive of a velocity gradient in the perpendicular direction $PA = 135°$, with an excess of redshifted emission toward the NW. Consistent with our interpretation of the NH$_3$ VLA-D data, this suggests inward motion in a rotationally-flattened flow that is seen not quite edge-on. The same redshifted feature is also marginally detected in the lower excitation CH$_3$CN transition as well as the OCS line. However, the infall signature in emission is only tentative, and a clearer indication of infall at small scales comes from the redshifted NH$_3$ absorption in the VLA-BnA data (see below).

In general, observational experience suggests that CH$_3$CN, along with NH$_3$, is a reliable tracer of high-density molecular mass and accretion flows [41, 158, 232, 234]. CH$_3$CN has recently been detected in the outflow of the nearby low-mass star

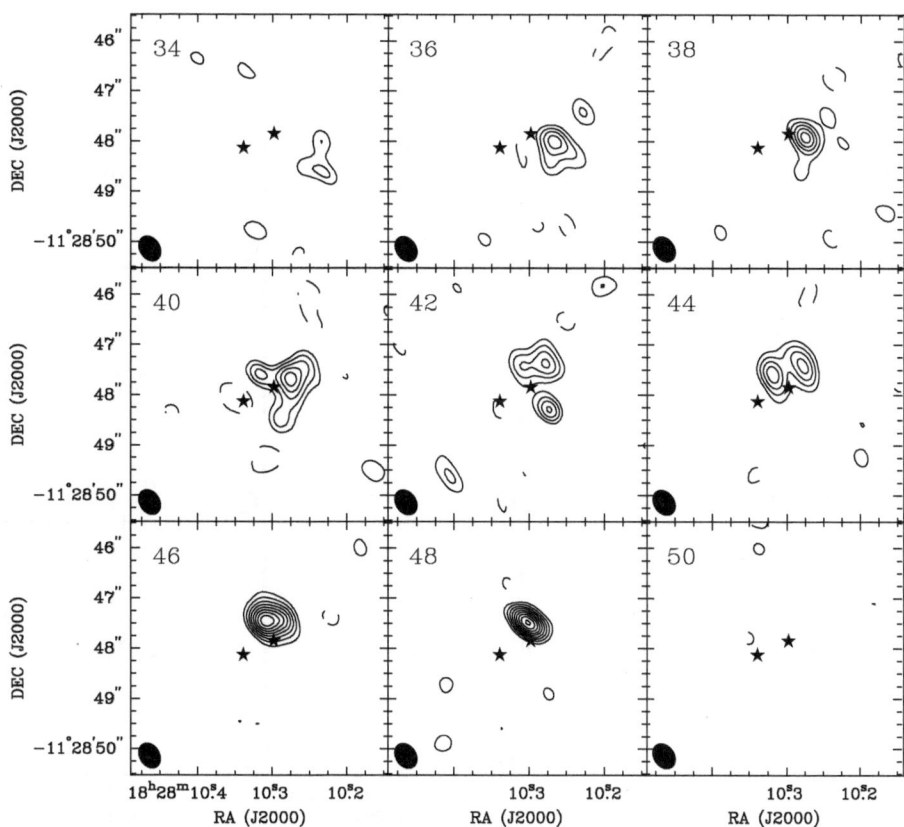

Fig. 3.8 Channel maps of SO_2 $J(K_a, K_b) = 11(1, 11) - 10(0, 10)$ from the SMA-VEX observations. Contours are $-4, 4, 6, 8, 10, 12, 14, 16, 18, 20 \times 30$ mJy beam^{-1} (negative in *dashed*, and positive in *solid*). The peak intensity is 612 mJy beam^{-1}

L1157 [10], but at a very low brightness (<0.03 K). The distribution of the OCS molecule in our observations is very similar to the CH_3CN, but the SO_2 velocities do not show the same pattern, and are more difficult to interpret. The SO_2 may be more easily affected by the excitation conditions, and part of the observed emission could arise from the shocked boundaries of outflows. From Gaussian fits to the CH_3CN $K = 2, 3$, and 4 emission lines at the position of the continuum peak, the systemic velocity at small scales is estimated to be $V_{sys} = 41.8 \pm 0.3$ km s^{-1}. Figure 3.14 shows the CH_3CN spectra and their Gaussian fits.

At subarcsecond angular resolution, our NH_3 observations are sensitive to emission of brightness temperature above ~ 200 K. Therefore the thermal NH_3 is detected only in absorption against the bright continuum. As in Sect. 3.4.2.2, a comparison between the NH_3 absorption line velocity and V_{sys} shows an inward velocity of ≈ 2 km s^{-1} in front of H II region A (Fig. 3.15). The NH_3 (2,2) absorption line in front of H II region B is redshifted by ~ 2.0 km s^{-1} with respect to V_{sys}, implying inward motion and accretion toward H II region B as well

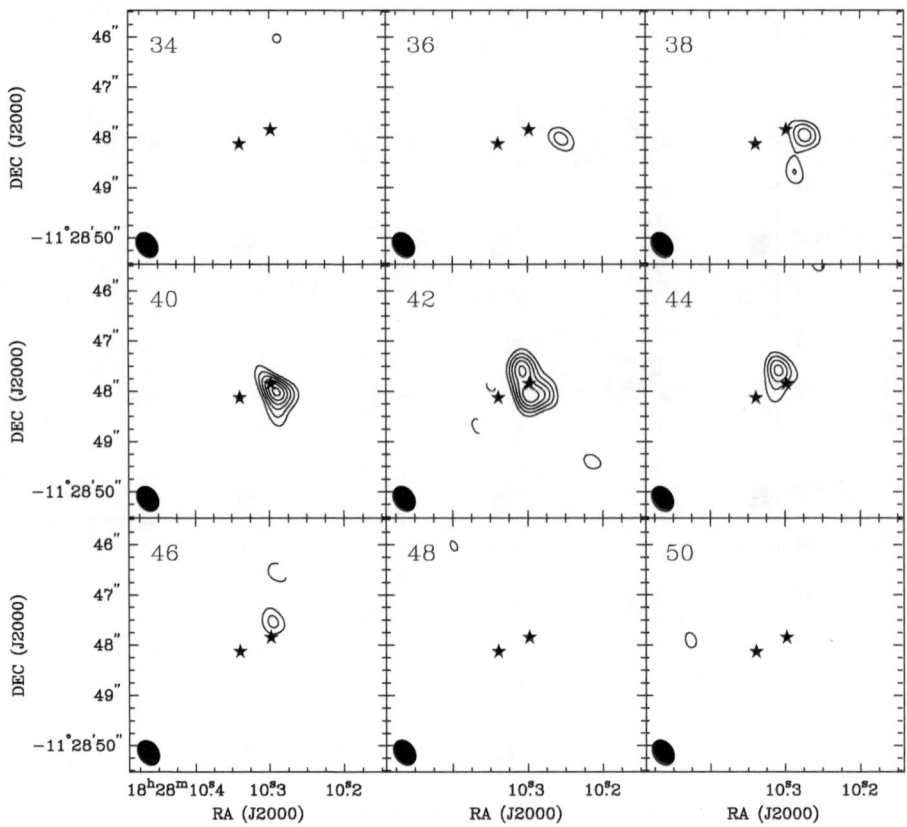

Fig. 3.9 Channel maps of CH$_3$CN $J(K) = 12(3) - 11(3)$ from the SMA-VEX observations. Contours are $-4, 4, 6, 8, 10, 12, 14 \times 30$ mJy beam^{-1} (negative in *dashed*, and positive in *solid*). The peak intensity is 440 mJy beam^{-1}

(Fig. 3.15). The (3,3) absorption in front of H II region B is mixed with NH$_3$ maser emission (Sect. 3.4.6), and the determination of the inward velocity is uncertain. More sensitive observations are needed to constrain the properties of the molecular gas around H II region B.

3.4.3.2   Core Parameters

We estimate the dynamical mass M (gas plus stars) within the smaller accretion flow in the same way as with the large-scale flow (see Sect. 3.4.2.2). At a radius $R \sim 5,000$ AU the rotation velocity is $V_{\rm rot} \sim 3\text{–}4$ km s^{-1}. Therefore, $M \sim 50\text{–}90\,M_\odot$. This is consistent with the lower limit to the stellar mass, $M_\star \approx 35\,M_\odot$, required for ionization equilibrium (Sect. 3.4.1.2). The estimate is also consistent with the gas mass derived from the mm continuum once the free–free contribution has been properly subtracted, $M_{\rm gas} \sim 35\text{–}95\,M_\odot$ (Sect. 3.4.1.2). The mean H$_2$ density is $n_{\rm H_2} \sim 10^6$ cm^{-3}.

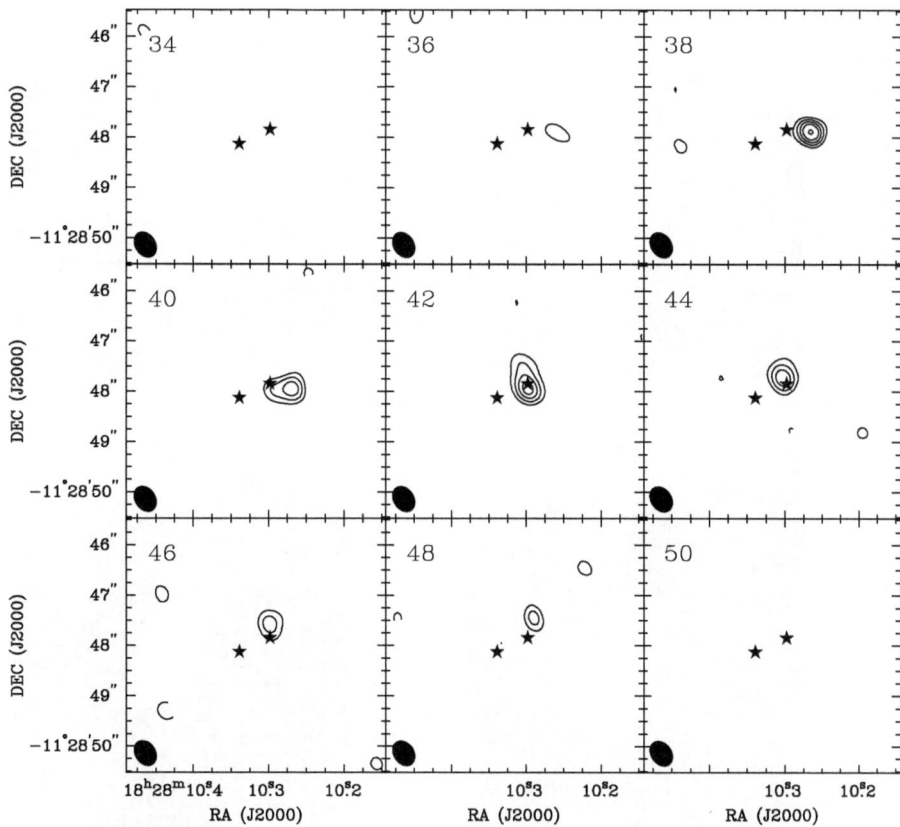

Fig. 3.10 Channel maps of CH_3CN $J(K) = 12(4) - 11(4)$ from the SMA-VEX observations. Contours are $-4, 4, 6, 8, 10, 12, 13 \times 30$ mJy beam^{-1} (negative in *dashed*, and positive in *solid*). The peak intensity is 401 mJy beam^{-1}

We derive the temperature in the dense gas surrounding H II region A from the rotational energy diagram of the lines of the CH_3CN $J = 12 - 11$ K-ladders (see e.g., Loren and Mundy [135] and Zhang et al. [232]). Figure 3.16 shows this diagram for two cases: one considering all the $K = 0, \ldots, 7$ lines, and the other including only the $K = 4, \ldots, 7$ lines, which have lower optical depths than the low-number K lines (also, the $K = 0, 1$ lines are blended, separated by only 5.8 km s^{-1}). The rotational temperature obtained for the former case is $T_{rot} \sim 403$ K, while for the latter it is $T_{rot} \sim 230$ K. The difference between the two values given above appears to be caused by optical depth effects (the rotational diagram analysis assumes that the emission is optically thin). Although our sensitivity level does not permit us to detect the lines of the isotopologue $CH^{13}CN$ and measure the optical depth of the CH_3CN emission, the upper limits are not restrictive ($\tau_{CH_3CN} < 20$). High optical

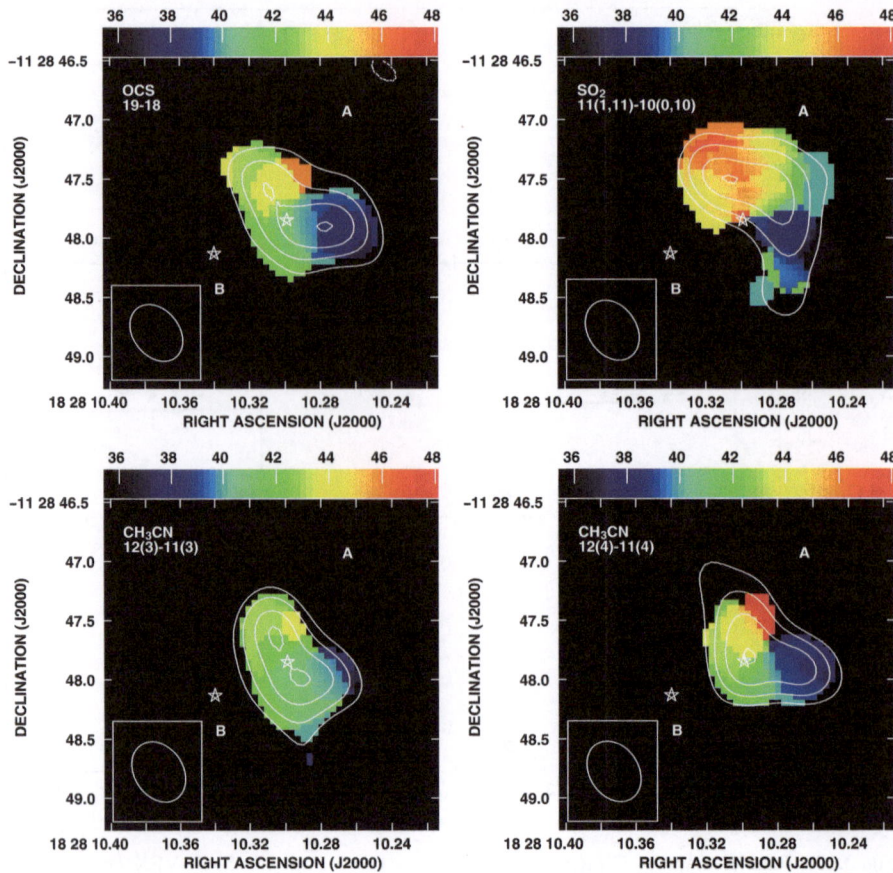

Fig. 3.11 Velocity-integrated emission (moment 0, *contours*) and intensity-weighted mean veloc-ity (moment 1, *color scale*) maps from hot-core molecules toward G20.08N. The *top left panel* shows the OCS $J = 19 - 18$ (contours are $-5, 5, 9, 13, 17 \times 0.15$ Jy beam^{-1} km s^{-1}). The *top right panel* corresponds to the SO$_2$ $J(K_a, K_b) = 11(1, 11) - 10(0, 10)$ (contours are $-5, 5, 8, 12, 15 \times 0.25$ Jy beam^{-1} km s^{-1}). The *bottom left panel* shows the CH$_3$CN $J(K) = 12(3) - 11(3)$ (contours are $-5, 5, 8, 12, 15 \times 0.15$ Jy beam^{-1} km s^{-1}). The *bottom right panel* plots the CH$_3$CN $J(K) = 12(4) - 11(4)$ (contours are $-5, 5, 9, 14, 19 \times 0.1$ Jy beam^{-1} km s^{-1}). Negative contours are in *dashed* style, and positive contours in *solid*. Only the continuum H II region A is associated with warm gas in our SMA-VEX data. The extent of the line emission is similar in all the tracers, although for the SO$_2$ it is more extended toward the northwest and more absorbed toward the continuum peak. The velocity gradient seen in the channel maps for each molecule is also seen here. The color scale is the same in all the frames

depths are also suggested by the flat slope of the K ladders (Fig. 3.1). Fitting the K ladders taking into account the opacities [166] yields a kinetic temperature close to the lower estimation, $T_k \sim 225$ K. The rotational temperatures are higher if we use

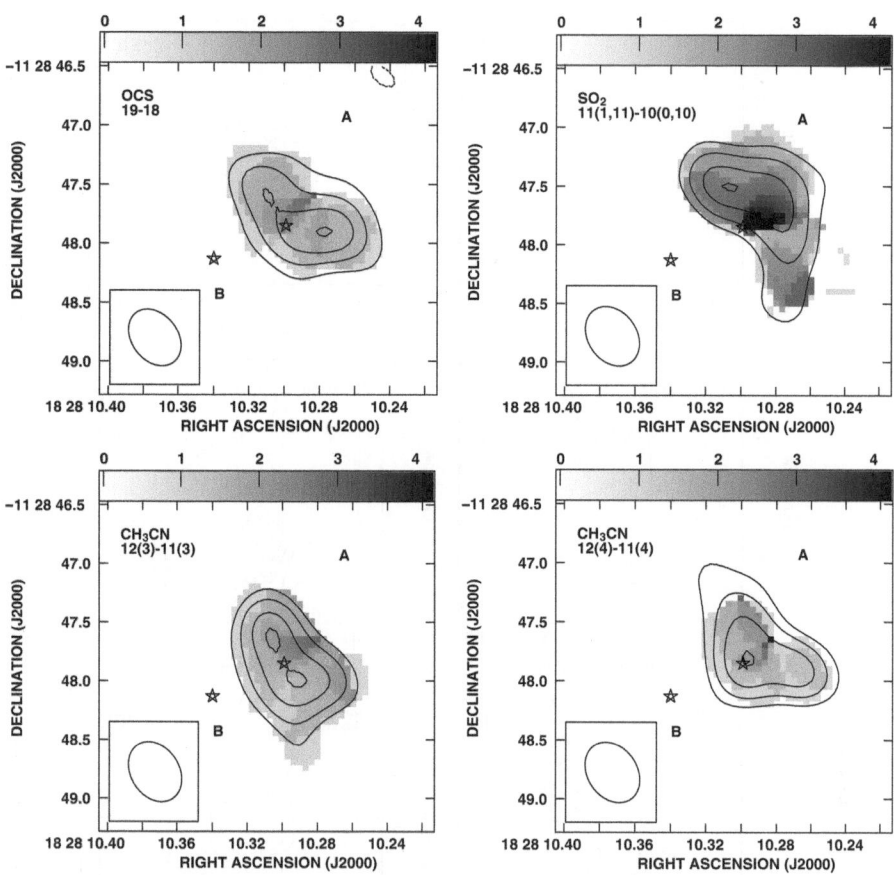

Fig. 3.12 Velocity-integrated emission (moment 0, *contours*) and velocity dispersion with respect to moment-1 velocity (moment 2, *gray scale*) maps from hot-core molecules toward G20.08N. Contours are the same as in Fig. 3.11. The plotted velocity dispersion is $\sigma = \mathrm{FWHM}/2(2\ln 2)^{1/2}$. σ increases toward the center of H II region A, possibly caused by unresolved motions toward the center

only the brightest pixels instead of averaging over all the emission. This suggests a temperature gradient toward the center of the H II region.

We derive a CH$_3$CN column density of $N_{\mathrm{CH_3CN}} \sim 7.5 \times 10^{15}\,\mathrm{cm}^{-2}$ assuming $T_k \sim 230\,\mathrm{K}$. Comparison of the CH$_3$CN column density with the dynamical mass implies an abundance $X(\mathrm{CH_3CN}) \sim 5 \times 10^{-9} - 2 \times 10^{-8}$ for the range of masses quoted above. Abundance estimates in other MSFRs cover a range of values: $\sim 10^{-10}$ inside the Orion hot core and $\sim 10^{-11}$ outside [135], 10^{-8} inside the Orion hot core and $10^{-9} - 10^{-10}$ in the Orion ridge [217], $\sim 3 \times 10^{-8}$ in Sgr B2(N) [154], $\sim 1 - 5 \times 10^{-7}$ in W51e1/e2 [168].

The optical depth of NH$_3$ in absorption toward H II region A is $\tau_{2,2} \sim 4.2$, $\tau_{3,3} \sim 0.9$. The mean column density is $N_{\mathrm{NH_3}} \sim 3.6 \times 10^{16}\,\mathrm{cm}^{-2}$. If the ammonia

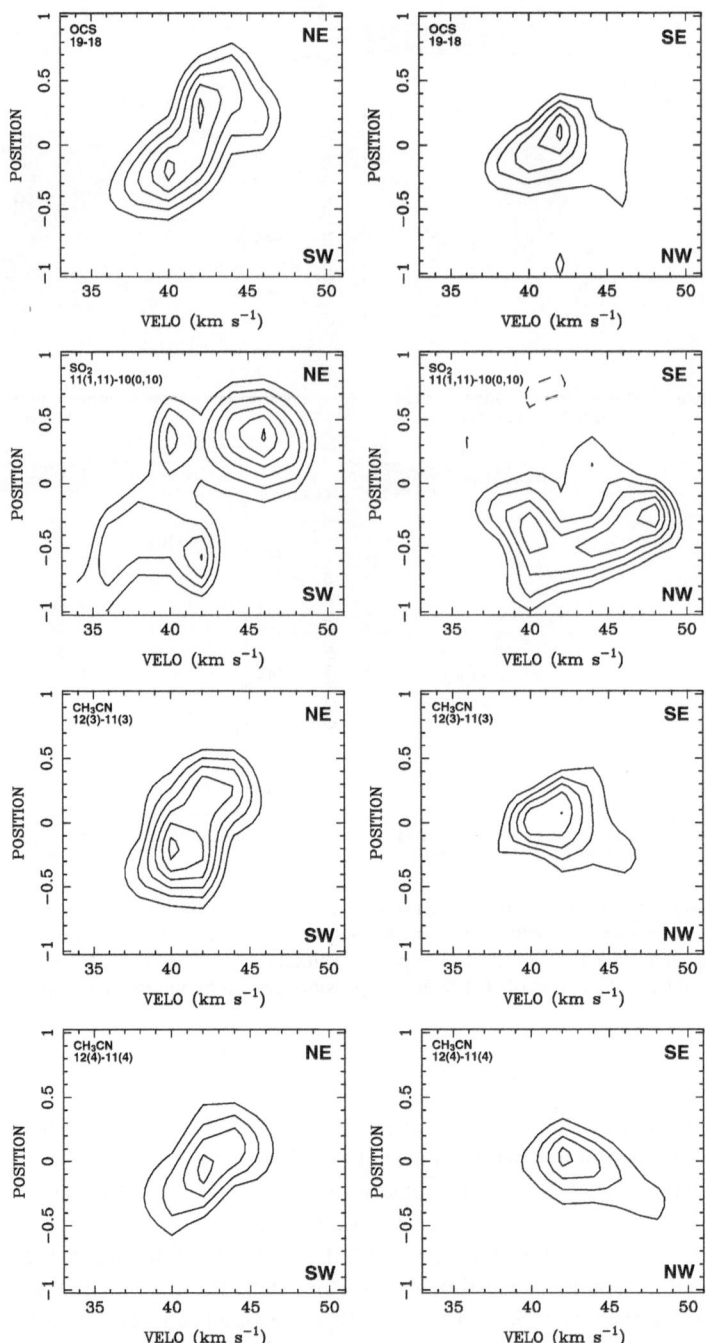

Fig. 3.13 Position-velocity diagrams of the SMA-VEX observations. The *left column* is for a cut at PA = 45° (SW–NE). The *right column* is at PA = 135° (NW–SE). Cuts are centered at the position of the continuum peak of H II region A. Contours are the same as in Fig. 3.7. The velocity gradient interpreted as rotation from SW to NE is clearly seen in OCS and CH$_3$CN. The SO$_2$ features are more complex

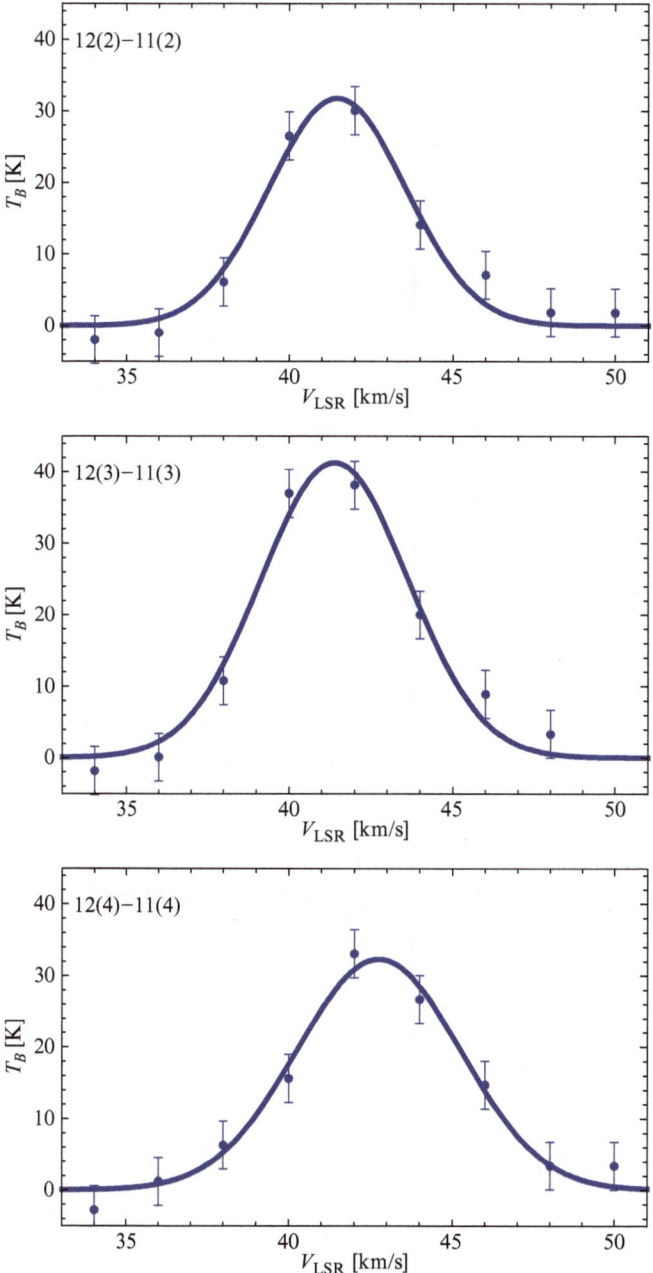

Fig. 3.14 Spectra (*points*) and Gaussian fits (*lines*) to the CH$_3$CN $J = 12 - 11$, $K = 2, 3$, and 4 lines at the position of H II region A. Error bars denote the 1σ noise in the 2 km s^{-1}-wide channels

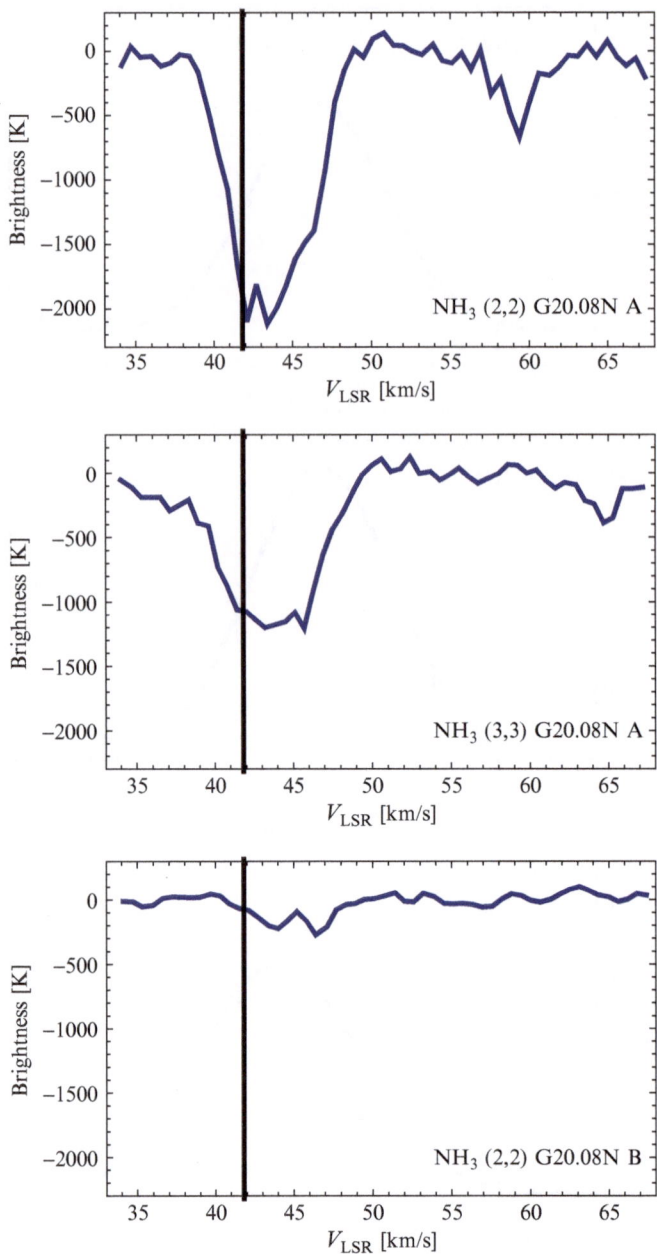

Fig. 3.15 *Top* and *middle panels*: NH₃ (2,2) and (3,3) spectra toward the absorption peak of H II region A in the VLA-BnA observations. The *vertical line* marks the systemic velocity ($V_{sys} = 41.8$ km s⁻¹) of the molecular gas at scales comparable to the H II region. The centers of the absorption lines are redshifted with respect to V_{sys}, indicating inflow of molecular gas toward H II region A at small scales. The (3,3) absorption is broader (FWHM ≈ 6.5 km s⁻¹) than the (2,2) (FWHM ≈ 5.6 km s⁻¹), probably caused by larger motions closer to the center. *Bottom panel*: NH₃ (2,2) spectrum toward H II region B. The absorption, considerably fainter than for H II region A, is also redshifted

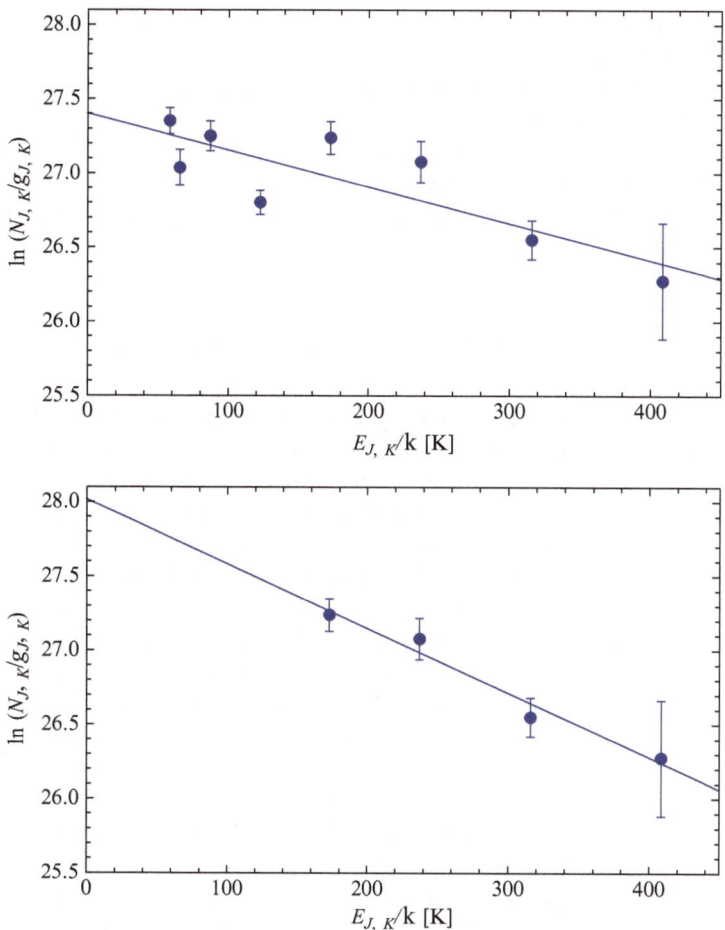

Fig. 3.16 Rotation diagram for the CH$_3$CN $J = 12 - 11$. The *top panel* is a fit of all the $K = 0, \ldots, 7$ lines. The *bottom panel* uses only the $K = 4, \ldots, 7$ lines, which have lower optical depths. Error bars are 3σ values. Lines are the linear fits to the data

abundance at these small scales is in the range $X(\mathrm{NH_3}) = 2 \times 10^{-6} - 1 \times 10^{-7}$, then the molecular hydrogen column density is $N_{\mathrm{H_2}} \sim 2 \times 10^{22} - 4 \times 10^{23}\,\mathrm{cm}^{-2}$. This implies a molecular gas mass of $M_{\mathrm{gas}} \sim 2\text{–}40\,M_\odot$. The kinetic temperature obtained from the NH$_3$ is $T_k \sim 50\,\mathrm{K}$, considerably cooler than that obtained for CH$_3$CN and implying that most of the NH$_3$ column density is further away from the H II region than the CH$_3$CN.

The distribution of the warm molecular gas around H II region A appears to be rotationally flattened (Figs. 3.7–3.12), but the observed size is slightly larger than the radius at which the accretion flow is expected to become centrifugally supported (i.e., a "disk"), $R_d = GM/V_{\mathrm{rot}}^2 \sim 2{,}000\text{–}3{,}000\,\mathrm{AU}$ for a star of mass $M_\star = 35\,M_\odot$

and $V_{rot} = 3\text{--}4$ km s^{-1}. Actually, the disk scale matches the size of the H II region (Table 3.3, Sect. 3.4.4).

The mass-inflow rate toward H II region A can be estimated from the high-resolution NH$_3$ absorption. From an inflow velocity of $V_{inf} \approx 2$ km s^{-1} and spherical geometry, the mass-inflow rate is $\dot{M} \sim 1 \times 10^{-3} - 2 \times 10^{-2} M_\odot$ year^{-1} (for the 2–40 $M_\odot$ of molecular gas detected in ammonia absorption). This estimate may be an upper limit because the hot molecular core is flattened rather than spherical.

3.4.4   The Ionized Gas: Radio Recombination Lines

In Sect. 3.4.1.2 we inferred a density gradient inside the HC H II region A. In this section, we derive the internal dynamics of this H II region based on multifrequency RRLs. The mm/sub-mm lines are especially important because they are much less affected by pressure broadening and preferentially trace denser gas. While subarcsecond resolution studies at wavelengths longer than 7 mm have been available for many years [56, 57, 176], similar studies at shorter wavelengths have had limited angular resolution (e.g., [95]). Keto et al. [114] presented the first results of high-frequency, high-resolution ($\sim 1''$), multifrequency RRL observations in a sample of 5 MSFRs with similar characteristics to G20.08N A. They were able to separate the contributions of pressure broadening and large-scale motions to the line width, even when the H II regions were unresolved. We follow their procedure to analyze our RRL data.

We observed the H30α ($\nu_0 = 231.90$ GHz) and H66α ($\nu_0 = 22.36$ GHz) lines at subarcsecond angular resolution (see Table 3.2). Because both the line-to-continuum ratio and the continuum intensity are lower at 22 GHz than at 231 GHz, the H66α line is much weaker than the H30α line. This is somewhat alleviated by the better sensitivity of the VLA, but the signal-to-noise ratio (S/N) in the high-frequency line is still better. Figure 3.17 shows the moment 0 and moment 1 maps of the H30α line. Although the emission is unresolved at half power, there is a slight indication of a velocity gradient in the ionized gas that agrees (not perfectly) with the rotation seen in CH$_3$CN and OCS. Figure 3.18 shows the H30α and H66α spectra toward H II region A and their Gaussian fits. The H66α line shows evidence of a blueshifted wing, suggesting either inflow or outflow in addition to rotation. Within the uncertainties of the fits, both the H66α and H30α lines have the same line width (Table 3.4).

Assuming that the dynamical broadening Δv_D (caused by turbulence and ordered motions) and the thermal broadening Δv_T are Gaussian, and that the pressure broadening Δv_L is Lorentzian, the RRL has a Voigt profile with line width [83]:

$$\Delta v_V(\nu) \approx 0.534 \Delta v_L(\nu) + (\Delta v_D^2 + \Delta v_T^2 + 0.217 \Delta v_L^2(\nu))^{1/2}, \qquad (3.3)$$

where all the widths are FWHM.

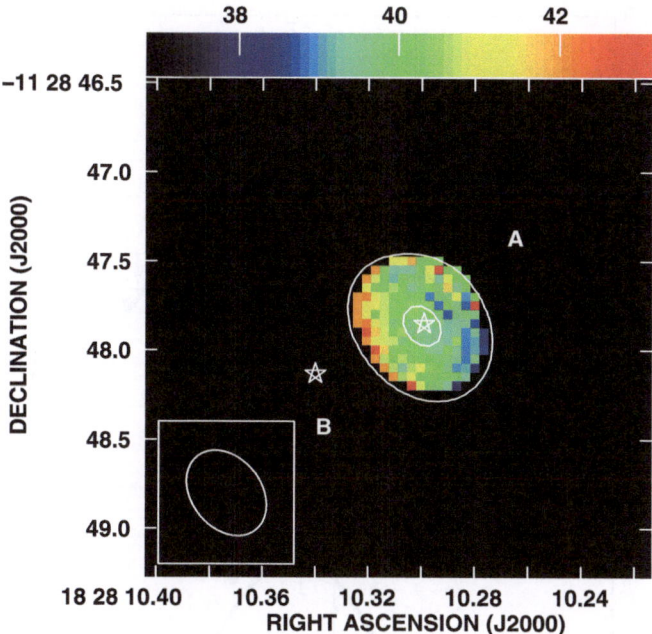

Fig. 3.17 Velocity-integrated emission (moment 0, *contours*) and intensity-weighted mean velocity (moment 1, *color scale*) map of the H30α RRL emission toward G20.08N. Contours are at $7, 50 \times 0.5$ Jy beam^{-1} km s^{-1}. Although the emission is unresolved at half power, the moment 1 map hints at the presence of a velocity gradient in the ionized gas similar to that seen in CH$_3$CN and OCS

For the H30α line at 231.9 GHz the pressure broadening is less than the thermal broadening at electron densities below 2×10^8 cm^{-3} [114]. Our SED modeling indicates lower densities over most of the HII region. Therefore, the observed line width can be attributed to thermal plus dynamical broadening. The electron temperature T_e in UC H II regions is typically $T_e = 8,000$–10,000 K, with a small gradient as a function of Galactocentric radius [4]. We adopt $T_e = 9,000$ K ($T_B <$ 100 K for H30α because of the low optical depth and $\lesssim 1$ filling factor), which translates into a thermal FWHM of $\Delta v_T = 20.9$ km s^{-1}. Therefore, from Eq. 3.3, we obtain a dynamical width of $\Delta v_D = 24.8$ km s^{-1}. From the velocity gradient (Fig. 3.17) it is seen that ~6 km s^{-1} of Δv_D can be in the form of rotation. The rest could be caused by inflowing or outflowing ionized gas, as suggested by the blueshifted (~2–3 km s^{-1}) mean velocities of the RRLs, and by the blue wing in the H66α spectrum (Fig. 3.18).

Most of the ionized gas that we see should be outflow. Inflow inside the H II region is expected within the radius where the escape velocity from the star exceeds the sound speed of the ionized gas. This is approximately the Bondi-Parker transonic radius [106], $R_b = GM_\star/2c_s^2 = 5.5$ AU $M_\star/M_\odot$, or about 190 AU for H II region A, assuming a sound speed $c_s = 9$ km s^{-1} and stellar mass $M_\star = 35\,M_\odot$. H II region A extends out to ~2,500 AU, so most of the gas is not gravitationally

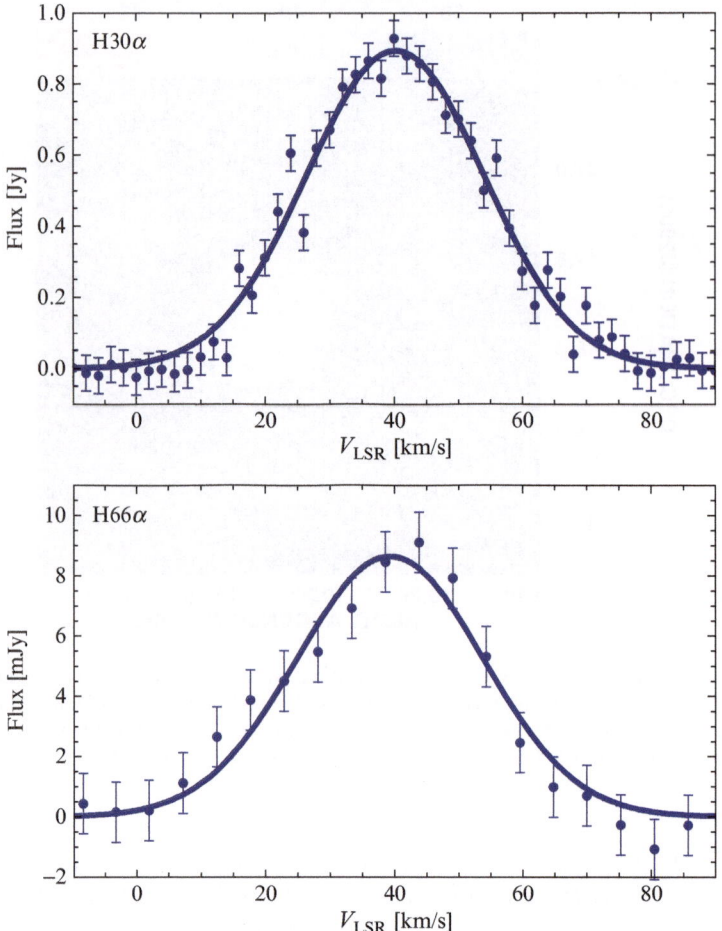

Fig. 3.18 Spectra (*points*) and Gaussian fits (*lines*) to the H30α (*top panel*) and H66α (*bottom panel*) lines toward G20.08N A. Error bars denote the 1σ noise in the channels. The channel spacing is 2 km s^{-1} for H30α and 5.2 km s^{-1} for H66α. The flux was integrated over a 0.5'' square box centered on H II region A

bound to the star and flows outward. In this model, the outflow is continuously supplied by photoevaporation off the rotationally flattened accretion flow. The somewhat misaligned velocity gradient in the ionized gas (Fig. 3.17) derives from a combination of the rotation and outflow blended together in the observing beam.

Table 3.4 Emission line parameters toward H II region A[a]

Species	Transition	V_{LSR} (km s^{-1})	FWHM (km s^{-1})	$T_{B,peak}$ (K)
H	66α	39.3 $\pm$ 1.1	34.1 $\pm$ 2.5	190 $\pm$ 13
H	30α	40.3 $\pm$ 0.4	32.5 $\pm$ 0.9	89 $\pm$ 2
CH$_3$CN	12(4)–11(4)	42.7 $\pm$ 0.3	5.8 $\pm$ 0.7	32 $\pm$ 4
CH$_3$CN	12(3)–11(3)	41.4 $\pm$ 0.2	5.3 $\pm$ 0.5	41 $\pm$ 4
CH$_3$CN	12(2)–11(2)	41.4 $\pm$ 0.2	4.9 $\pm$ 0.5	32 $\pm$ 3

[a]From Gaussian fits. 1σ statistical errors are quoted

3.4.5 Outflow Tracers

Low angular resolution (HPBW $= 14''$), single-dish (JCMT) observations of SiO (8–7) show evidence for large-scale molecular outflows in G20.08N [116]. Although the standard outflow tracers ^{12}CO $J = 2 - 1$ and ^{13}CO $J = 2 - 1$ are in our passband, we do not detect any CO in emission. The (u,v) coverage of our SMA-VEX observations is incapable of imaging structures larger than $\sim 4''$; therefore, the CO emission from the molecular cloud and molecular outflow must be on larger scales. ^{12}CO and ^{13}CO are seen in our data in absorption at the position of H II region A, at several different velocities in the range $V_{LSR} = 42$–84 km s^{-1}. Some CO absorption features are at the same velocities as the HI absorption features of Fish et al. [65] and are therefore due to foreground gas that is not related to G20.08N but rather to intervening Galactic spiral arms.

3.4.6 A New NH$_3$ (3,3) Maser

A handful of NH$_3$ masers have been reported in the literature, always associated with massive star formation (e.g., [211]). Many of the known NH$_3$ masers are from non-metastable $(J > K)$ transitions. The first clear detection of a metastable $(J = K)$ NH$_3$ (3,3) maser was reported by Mangum and Wootten toward DR 21(OH) [140]. Most of the detections point toward a shock excitation origin for the population inversion, inasmuch as the maser spots are invariably associated with outflow indicators such as bipolar CO and/or SiO structures, class I methanol masers, and/or water masers [118, 140, 229].

We report the serendipitous detection of a new NH$_3$ (3,3) maser toward G20.08N. The maser spot is relatively weak, and is spatially centered at α(J2000) $=$ 18^h 28^m 10^{s}346, δ(J2000) $= -11° 28' 47''.93$, close in projection to the center of H II region B. The maser spot is spatially unresolved even in uniform weighting maps of the VLA-BnA data (HPBW $= 0.48'' \times 0.26''$, PA $= -0.8°$). If the deconvolved source size is limited to half the beam size, then the peak brightness temperature of the spot is constrained to $T_B > 7 \times 10^3$ K. The high intensity, together with the absence of similar emission in our (2,2) maps at high angular resolution,

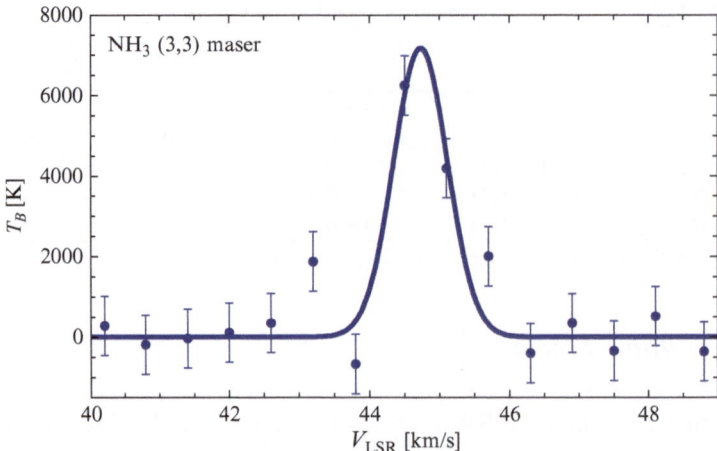

Fig. 3.19   VLA-BnA spectrum of the NH$_3$ (3,3) maser spot toward G20.08N. The brightness temperature T_B scale assumes an angular size equal to half the uniform-weighted beam dimensions

confirm the maser nature of the (3,3) emission. The spectral feature is also very narrow (Fig. 3.16), typical of maser emission, although it shows evidence of line wings. From a Gaussian fit to the line profile, the velocity of the maser is $V_{maser} \sim 44.7$ km s^{-1} (accurate only to 20–40%). The FWHM is $\sim$0.7 km s^{-1}, after deconvolving the channel width of 0.6 km s^{-1}. Owing to its position, it is probable that the maser is excited by H II region B. We do not have sufficient data, however, to assert that it is excited in a shock (Fig. 3.19).

3.5   Hierarchical Accretion in G20.08N

3.5.1   The Observations

One of the central questions in star formation is whether star formation is "bimodal", i.e., whether high-mass and low-mass stars form in a different way [184]. Analysis of recent observations suggests that accretion flows around protostars of up to $M_\star \sim 10 M_\odot$ can be explained within the standard model of star formation as a combination of a thin disk inside a rotationally flattened envelope [109, 123, 147]. Of course, once a protostar gains the mass and temperature of an O star, the formation of an H II region within the accretion flow introduces new phenomena [104, 106, 108, 161]. Yet not all molecular clouds form massive stars. In a comprehensive survey, Solomon et al. [191] found that clouds that form low-mass stars are uniformly distributed throughout the Galactic disk, but those that form the clusters of the most massive stars, O stars capable of producing significant H II regions, are associated with the galactic spiral arms. If the formation process is very similar (with the addition of significant ionization for $M_\star > 20 M_\odot$), then the difference may be found in the conditions of the parent molecular clouds. The observations of G20.08N reported in this chapter and of G10.6–0.4 reported previously [103, 108] suggest that one difference is that the molecular clouds surrounding young clusters of O stars are in a state of overall collapse whereas in star-forming regions without H II regions we see only localized collapse.

In G20.08N the H II regions A, B, and C are surrounded by a common molecular cloud of radius ~ 0.5 pc and mass $M \sim 2,000 M_\odot$, which is rotating with a velocity of ~ 4 km s^{-1} and contracting with an inward velocity of ~ 2 km s^{-1}. This rotation and contraction constitute a large-scale accretion flow. The velocity of the inward flow is about equal to the rotational velocity implying that the gas is approximately in free-fall and not constrained by centrifugal force. Within this larger flow are at least two smaller accretion flows around H II regions A and B. The molecular core around H II region A is bright enough to be detected in hot-core molecules in emission at subarcsecond angular resolution. The radius and mass of this core are 0.05 pc and $\sim 20 M_\odot$. Accretion in the core is indicated by rotation at a velocity of 3 or 4 km s^{-1} and contraction of ~ 2 km s^{-1}. We do not detect molecular emission around H II region B, but the accretion inflow is inferred from NH$_3$ absorption that is redshifted by ~ 2 km s^{-1} with respect to the systemic velocity.

In contrast, star-forming regions that contain only low mass stars or even stars as massive as type B do not appear to have this global collapse of the entire parent cloud. The observations of these regions suggest only localized accretion flows within individual clumps. For example, star-forming regions such as IRAS 19410+2336 [28, 29, 195], IRAS 05358+3543 [27, 131], and AFGL 5142 [234, 235], have bolometric luminosities of at most a few times $10^4 L_\odot$, consistent with type B protostars. There are a number of cores of size similar to those in G20.08N, but there is no reported evidence of a larger parsec-scale, accretion flow.

In an analysis of a recent numerical simulation, Vázquez-Semadeni et al. [209] also find that the formation of massive stars or clusters is associated with large-scale collapse involving thousands of $M_\odot$ and accretion rates of $10^{-3} M_\odot$ yr^{-1}. In contrast, low- and intermediate-mass stars or clusters in their simulation are associated with isolated accretion flows that are a factor of 10 smaller in size, mass, and accretion rate.

If global collapse of the host molecular cloud is necessary for the formation of O stars in clusters, then the association of O stars with galactic spiral arms may imply that compression of giant molecular clouds as they pass through galactic spiral arms may be the primary mechanism for initiating global collapse [143, 170, 183]. The low-mass star-forming regions found by Solomon et al. [191] to be spread throughout the Galaxy may not need such large-scale compression.

3.5.2 Resupply of the Star-Forming Cores

The orientations of the large cloud-scale accretion flow and the core-scale flow around H II region A are similar and the flows could be continuous (we do not know the orientation of the flow around H II region B because we cannot detect the surrounding molecular emission). The molecular core around H II region A contains only a few tens of $M_\odot$, similar to the mass of a single O star. If the core is to form one or more O stars at less than 100% efficiency, its mass must be resupplied by the larger scale accretion flow. Resupply is also suggested by the short dynamical or crossing timescale, given by the ratio of the size to the infall velocity, $\sim 10^4$ years. If the core is to last more than this, it must be resupplied by the larger scale flow, which has a dynamical timescale of $\sim 10^5$ years. If the core is not resupplied then the growing protostar may simply run out of gas before reaching the mass of an O star.

A similar process of resupply is suggested in recent theoretical work. Analytic arguments show that as an unstable cloud fragments, there should be a continuous cascade of mass from larger to smaller fragments as well as a cascade of kinetic energy [64, 153]. In recent numerical simulations of high mass star formation, Wang et al. [212] find that most of the mass is supplied from outside a 0.1 pc core around the protostar. Vázquez-Semadeni et al. [209] and Peters et al. [161] also find in their simulations that as a massive core is consumed from the inside by an accreting protostar, the core continues to accrete mass from the outside.

3.5.3 Accretion Rate

The accretion rate within the small-scale flow around H II region A is $\dot{M} \sim 1 \times 10^{-3}$ to $2 \times 10^{-2} M_\odot$. Similar accretion rates are reported for flows around other H II regions such as G10.6–0.4, $8 \times 10^{-4} M_\odot$ year^{-1} [103], and W51e2, 3×10^{-3} to $1 \times 10^{-2} M_\odot$ year^{-1} [226]. In contrast, the accretion rates estimated for the cores in these MSFRs without bright H II regions are generally lower by one or two orders of magnitude, $\sim 10^{-4} M_\odot$ yr^{-1} [27, 28].

A high accretion rate is necessary to form an O star. Because accreting massive protostars begin core hydrogen burning well before reaching the mass of an O star, they evolve essentially as main-sequence stars of equivalent mass. Numerical simulations of stellar evolution that include accretion predict that unless the accretion rate is high enough, a growing massive protostar will evolve off the main-sequence and explode before it reaches the mass of an O star [108]. For a protostar to gain the mass of an O star, the rate at which accretion supplies fresh hydrogen to the growing protostar must be greater than the rate at which the star burns the hydrogen. At the upper end of the mass spectrum, $M_\star > 40 M_\odot$, this rate is $> 10^{-3} M_\odot$ yr^{-1}. The accretion flows in G20.08N are capable of supplying gas at the rate necessary to form massive O stars. In MSFRs without O stars, the accretion rates may be too low for the stars to achieve the mass of an O star within their hydrogen-burning lifetimes.

3.5.4 Transfer of Angular Momentum

If the flow in G20.08N is continuous, then the observations show that the flow spins up as it contracts. Ignoring projection effects, the magnitude of the specific angular momentum on the large scale is $L/m \sim 0.6$ km s^{-1} pc at $r \approx 0.2$ pc, from the VLA-D NH$_3$ data, while from the VEX SMA data it is $L/m \sim 0.1$ km s^{-1} pc at $r \approx 0.02$ pc around H II region A. From these estimates $\sim$85% of the specific angular momentum in the large-scale flow is lost. In previous observations of G10.6–0.4 [103], it was found that 97% of the angular momentum in that flow is lost between 1.5 and 0.02 pc. Evidently, angular momentum is efficiently transferred outward and does not prevent collapse of the cloud.

Magnetic fields may be important in this process [81]. Although there are no observations of the magnetic field in G20.08N, the field direction has been mapped in another MSFR with HC H II regions, W51e1/e2. Here, observations show that the magnetic field is uniform on the larger scale, 0.5 pc [129], while on the smaller scale, 0.03 pc, of the accretion flow onto W51e2, the field is pinched into an hourglass shape with the accretion flow at the waist [197]. Thus ordered field lines extend from the H II region-scale accretion flow to the large-scale molecular cloud, and if the field has enough strength, angular momentum could be transferred outward by the field. However, dust polarization observations do not give a direct estimate of the field strength, and it is also possible that the field is essentially passive and just carried along by the flow. The observation that the clouds in both W51e1/e2 and G20.08N are close to free-fall collapse implies that the magnetic field is not strong enough to support the clouds. In other words, these clouds are magnetically supercritical.

Numerical simulations of star formation that do not include magnetic forces show that angular momentum can be transferred by hydrodynamics alone. Abel et al. [1] find that at any radius, there is both low and high angular momentum gas, and that pressure forces or shock waves can redistribute the angular momentum between fluid elements. Lower angular momentum gas sinks inward and displaces higher angular momentum gas outward, resulting in a net outward flow of angular momentum.

Whether the specific angular momentum is transferred by hydrodynamics or magnetic forces, observations and simulations show that angular momentum is not conserved as a function of radius, and does not prevent the gas from flowing continuously from large to small scales in a rotating flow.

3.6 Conclusions

We report radio and mm observations of the molecular and ionized gas toward the O-star cluster G20.08N, made with an angular resolution from $\sim$0.1 to $\sim$0.01 pc. Our main findings can be summarized as follows:

1. We find a large-scale ($\sim$0.5 pc) accretion flow around and into a star cluster with several O-type stars, identified by one UC and two HC H II regions. This flow is rotating and infalling towards its center. The two HC H II regions are surrounded by smaller accretion flows ($\sim$0.05 pc), each of them with the signature of infall too. The brightest (toward H II region A) is detected in mm emission lines, and rotates in concordance with the large-scale flow.

2. The similar orientations of the flows at small and large scales, as well as their dynamical timescales ($\sim10^4$ and $\sim10^5$ years respectively), and masses ($\sim10M_\odot$ and $\sim10^3\,M_\odot$ respectively), suggest that, if O stars are forming in G20.08N (as it is observed), then the smaller scales ought to be resupplied from the larger scales. The same result has been found in recent numerical simulations of massive star formation in clusters.

3. The brightest HC H II region (A) has a rising SED from cm to mm wavelengths and broad hydrogen recombination lines. Both characteristics suggest density gradients and supersonic flows inside the H II region. A velocity gradient is tentatively detected in the recombination line emission of this source, suggesting rotation and outflow in the ionized gas at the innermost scales. H II region A can be interpreted as the inner part of the surrounding molecular accretion flow, with the observed ionization being produced by photoevaporation.

Chapter 4
Time Variability of HII Regions: A Signature of Accretion?

4.1 Summary

Over a timescale of years, an observed change in the optically thick radio continuum flux can indicate whether an unresolved H II region around a newly formed massive star is changing in size. In this chapter we present a study of archival VLA observations of the hypercompact H II region G24.78 + 0.08 A1. We found a decrease of ~45% in its 6-cm flux over a 5 year period. Such a decrease indicates a contraction of ~ 25% in the ionized radius and could be caused by an increase in the ionized gas density if the size of the H II region is determined by a balance between photoionization and recombination. This finding is not compatible with continuous expansion of the H II region after the end of accretion onto the ionizing star, but is consistent with the hypothesis of gravitational trapping and (partially) ionized accretion flows if the mass-accretion rate is not steady.

These results have been published in Galván-Madrid, Roberto, Rodríguez, Luis F., Ho, Paul T. P., and Keto, Eric R. "Time Variation in G24.78 + 0.08 A1: Evidence for an Accretion Hypercompact H II Region?" The Astrophysical Journal Letters, 674, L33 (February 2008) [73].

4.2 Introduction

The formation of massive stars ($M_\star > 8\,M_\odot$) by accretion presents a number of theoretical difficulties; among them, that once a star attains a sufficient mass its surface temperature is high enough to produce a small H II region (see the reviews on ultracompact and hypercompact H II regions by Churchwell [46]; Kurtz [124]; and Hoare et al. [90]). The thermal pressure differential between the hot ($\sim 10^4$ K) ionized gas and the cold (~ 100 K) molecular gas can potentially reverse the accretion flow of molecular gas and prevent the star from ever reaching a higher mass. However, models that include the effects of gravity [105] have shown that

R.J. Galván-Madrid, *On the Formation of the Most Massive Stars in the Galaxy,*
Springer Theses, DOI 10.1007/978-1-4614-3308-8_4,
© Springer Science+Business Media New York 2012

if the H II region is small enough (ionized diameter $\leq 1,000\,\mathrm{AU}$ for total stellar mass $\geq 350\,M_\odot$), the gravitational attraction of the star(s) is greater than the thermal pressure of the ionized gas, permitting the molecular accretion flow to cross the ionization boundary and proceed toward the star as an ionized accretion flow within the H II region. In this stage the H II region is said to be gravitationally trapped by the star. Of course, geometry does not need to be spherical, rather, the accretion flow is expected to be flattened. Models that take into account the effects of the geometry of accretion show that the H II region can be outflowing in the polar directions, where the parental accretion flow is less dense, while it can be gravitationally trapped along the plane of the accretion flow (disk).

Previous studies have demonstrated a number of observational techniques bearing on the evolution of small H II regions around newly formed stars. Accretion flows onto and through H II regions can be directly observed by mapping molecular and radio recombination lines (RRLs) at very high angular resolution [108, 189]. If the H II region is too small to be spatially resolved, the frequency dependence of the velocities and widths of RRLs can be used to infer steep density gradients and supersonic velocities within the H II regions [114], as expected from the presence of ionized accretion flows or bipolar outflows. Radio continuum observations at very high angular resolution, made at two epochs some years apart, can also be used to directly observe changes in size of small H II regions [67]. Changes in size indicate whether the evolution of the H II region is consistent with pressure-driven expansion or gravitational trapping.

In this chapter we present a study that demonstrates yet another technique – how a comparison of radio continuum observations at different epochs can be used to infer size changes in H II regions even if the observations do not spatially resolve them. At optically thick frequencies the radio continuum flux depends to first approximation only on the size of an H II region and is independent of its internal density structure. Therefore, even if the size is not known, a change in the flux over time still indicates a change in the size of the H II region. This technique is particularly valuable because it can be used on the smallest and youngest H II regions, and because lower angular resolution radio continuum observations generally require less observing time than spectral line and high angular resolution observations.

For this study we selected the hypercompact (HC) H II region G24.78 + 0.08 A1 which lies at the center of a massive molecular accretion flow [20, 22] and also has multi-epoch 6 cm radio continuum observations in the VLA archive. G24.78 + 0.08 was detected in the centimetric (cm) continuum by Becker et al. [19], and later resolved by Codella et al. [47] into a compact (A) and an extended (B) component. A millimeter (mm) interferometric study [20] revealed the presence of two massive rotating toroids centered in respective dust cores (A1 and A2). The compact cm emission comes from the mm component A1 (hereafter G24 A1), and has recently been resolved by Beltrán et al. [23]. If G24 A1 is ionized by a single star, its spectral type should be earlier than O9 [47]. Also, G24 A1 likely powers a massive CO outflow [70].

The infalling and rotating molecular gas and bipolar outflow all suggest ongoing accretion. However, based on proper motions of H_2O masers around the H II region,

Beltrán et al. [23] and Moscadelli et al. [149] proposed that at the present time, the H II region is expanding into the accretion flow. The suggested timescale for the expansion is short enough that we should be able to detect a corresponding increase in the optically thick radio continuum flux within a few years.

4.3 Observations

We searched the VLA[1] archive for multi-epoch observations centered in the G24 A1 region at optically thick frequencies (below 23 GHz for G24 A1). Since in the optically thick part of the spectrum the flux density of the source scales as the angular size squared, flux density variations corresponding to size variations can be detected even in observations of modest angular resolution. We chose three data sets of 6-cm observations in the C configuration, two from 1984 and one from 1989 (see Table 4.1).

The observations were made in both circular polarizations with an effective bandwidth of 100 MHz. The amplitude scale was derived from observations of the absolute amplitude calibrator 3C286. This scale was transferred to the phase calibrator and then to the source. We estimate an error not greater than 10% for the flux densities of the sources.

We edited and calibrated each epoch separately following the standard VLA procedures using the reduction software AIPS. Precession to J2000 coordinates was performed running the task UVFIX on the (u,v) data. After self-calibration, we made CLEANed images with uniform weighting and cutting the short spacings (up to 10 Kλ) to minimize the presence of extended emission at scales larger than $\sim 20''$.

Before subtraction, we made the images as similar as possible. We restored the CLEAN components with an identical Gaussian beam HPBW $4\overset{''}{.}79 \times 3\overset{''}{.}38$, PA$=-15°$. We applied primary beam corrections and aligned the maps. No significant differences were found in the difference image between the 1984 May 11 and 14 images, as expected for such a small time baseline. We therefore averaged the two 1984 epochs into a single data set. In subtracting the final maps we allowed

Table 4.1 Observational parameters

Epoch	Phase center[a] α(J2000)	δ(J2000)	Amplitude calibrator	Phase calibrator	Bootstrapped flux density (Jy)	Beam (arcsec × arcsec; deg)
1984 May 11	18 36 12.145	−07 11 28.17	3C286	1743−038	2.414 ± 0.005	5.38 × 3.44; −29
1984 May 14	18 36 12.145	−07 11 28.17	3C286	1743−038	2.464 ± 0.007	4.78 × 3.34; +03
1989 Jun 23	18 36 10.682	−07 11 19.87	3C286	1834−126	0.153 ± 0.001	4.50 × 3.31; −10

[a]Units of right ascension are hours, minutes, and seconds. Units of declination are degrees, arcminutes, and arcseconds

---

[1]The National Radio Astronomy Observatory is operated by Associated Universities, Inc., under cooperative agreement with the National Science Foundation.

the 1984 image to have small ($\simeq 0.1''$) shifts in position as well as a scaling of $\simeq 10\%$ in amplitude. This was done in order to minimize the rms residuals of the difference image in the region of interest. A similar procedure was used by Franco-Hernández and Rodríguez [67] to detect a variation in the lobes of the bipolar UC H II region NGC 7538 IRS1. The individual maps, as well as the final difference image between 1984 and 1989 are shown in Fig. 4.1.

4.4 Discussion

4.4.1 The Expected Variation Trend

4.4.1.1 If G24 A1 Is Expanding

Beltrán et al. [23] suggested that the 7 mm and 1.3 cm continuum morphologies are consistent with limb-brightening from a thin, ionized-shell structure. Based on H_2O maser proper motions [23, 149] they also suggested an expansion speed of $\sim$40 km s^{-1}. The increase in flux corresponding to the increase in size due to expansion ought to be detectable in a few years.

4.4.1.2 If G24 A1 Is Accreting

If G24 A1 is the ionized inner portion of the star forming accretion flow, the long-term growth of the H II region due to the increasing ionizing flux of the star should be imperceptible. However, the H II region could change in size over an observable timescale if the gas density in the accretion flow is time variable. Because the mass of ionized gas within the H II region is very small compared to the mass of the accretion flow, even a small change in the flow density could affect the size of the H II region.

For example, we can obtain a lower limit for the ionized mass of G24 A1 M_{HII} from a measurement of its flux density at a frequency for which its emission is optically thin, and assuming that it is spherical and homogeneous. From the equations of Mezger and Henderson [146] we have that:

$$\left[\frac{M_{HII}}{M_\odot}\right] = 3.7 \times 10^{-5} \left[\frac{S_\nu}{mJy}\right]^{0.5} \left[\frac{T_e}{10^4\,K}\right]^{0.175} \times \left[\frac{\nu}{4.9\,GHz}\right]^{0.05} \left[\frac{D}{kpc}\right]^{2.5} \left[\frac{\theta_s}{arcsec}\right]^{1.5},$$

$$(4.1)$$

where the flux density is S_ν, T_e is the electron temperature, D is the distance to the region, and θ_s is its FWHP. Taking $S_\nu = 101$ mJy, $D = 7.7$ kpc, and $\theta_s = 0''17$ at 7 mm [23], and assuming $T_e = 10^4$ K, the ionized mass in G24 A1 would be $M_{HII} \sim 5 \times 10^{-3} M_\odot$. A higher limit to the mass can be obtained considering that HC

Fig. 4.1 VLA images of G24.78 + 0.08 for 1984.36 (*top*), 1989.48 (*middle*), and the difference of 1989.48 – 1984.36 (*bottom*). The contours are −10, −8, −6, −5, −4, 4, 5, 6, 8, 10, 12, 15, 20, 30, 40, and 60 times 0.57 mJy beam^{-1}. The half power contour of the synthesized beam (4.''79 × 3.''38 with a position angle of −15°) is shown in the *bottom left corner* of the images. The crosses indicate the positions of the components A1 and B from our Gaussian fits to the 1984.36 image. The negative residuals observed in the difference image indicate a decrease of ∼45% in the flux density of component A1

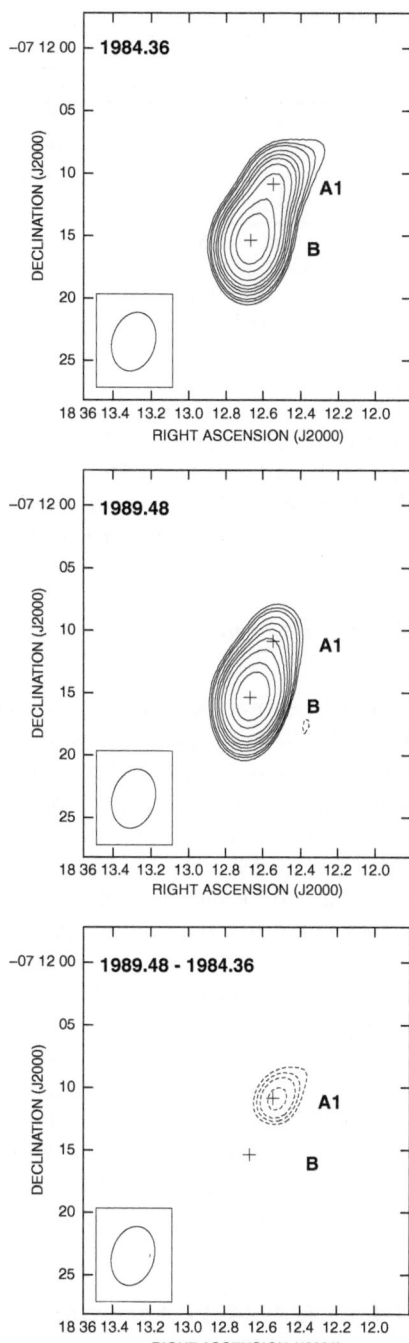

Table 4.2 Gaussian fit results

Epoch	Component	Position[a,b] α(J2000)	δ(J2000)	Flux density (mJy)
1984.36	A1	18 36 12.545	−07 12 10.87	11.4±0.8
1984.36	B	18 36 12.668	−07 12 15.37	31.4±1.0
1989.48	A1	18 36 12.545	−07 12 10.87	06.3±0.8
1989.48	B	18 36 12.668	−07 12 15.37	32.8±0.8

[a] Units of right ascension are hours, minutes, and seconds. Units of declination are degrees, arcminutes, and arcseconds.
[b] For the 1989.48 epoch, the centers of the Gaussians were fixed to those obtained for the 1984.36 epoch

H II regions should have density gradients ($n \propto r^{\alpha}$, with $\alpha = -1.5$ to -2.5) rather than being homogeneous [106]. In this case we obtain ionized masses between $M_{HII} \sim 1 \times 10^{-2} \, M_{\odot}$ and $M_{HII} \sim 3 \times 10^{-2} \, M_{\odot}$, a factor of $2 - 6$ higher than under the assumption of homogeneity, but still very small when compared to that of the surrounding molecular material or the ionizing star itself.

4.4.2   The Observed Variation

Figure 4.1 shows the individual maps of the 1984 May $11 + 14$ (1984.36) and 1989 June 23 (1989.48) epochs, as well as the difference image. Sources A1 and B (the former labeled A by Codella et al. [47]) are unresolved ($\theta_s \leq 2''$), and the total emission is dominated by component B. However, the flux decrease is centered at the position of A1. We performed Gaussian fits to both components using the task JMFIT in AIPS, and the results are summarized in Table 4.2. We have checked the reliability of the fits to the image by making direct fits to the (u,v) data. The values obtained from both techniques are entirely consistent but suggest that the errors given by the tasks used in the image (JMFIT) and (u,v) (UVFIT) fittings are underestimated by a factor of 2. The errors given in Table 4.2 have been corrected by this factor.

The decrease in the flux density of G24 A1 between 1989.48 and 1984.36 is 5.1 ± 1.1 mJy, or $45 \pm 10\%$. Component B shows no evidence of time variability, as expected for a more evolved H II region not associated with signs of current star-forming activity such as H_2O or OH masers [47]. We set an upper limit of 2% to the circularly-polarized emission of the sources, which indicates that we are not dealing with variable gyrosynchrotron emission from an active stellar magnetosphere.

4.4.3   Is G24 A1 Accreting?

The $\sim$45% flux decrease (i.e., a contraction of $\sim$25% in the ionized radius) we have detected at 6 cm toward G24 A1 is not consistent with the hypothesis that this H II region is expanding rapidly into the molecular accretion flow. Assuming a radius

$R \sim 500$ AU and an expansion velocity $v \sim 40$ km s^{-1} [23], a flux *increase* of $\sim 20\%$ should have been observed between the compared epochs.

We attribute the contraction of the H II region to an increase in its density produced by the enhancement of accretion, either caused by an isotropic increment in the mass-accretion rate or by the sudden accretion of a localized clump in the neutral inflow (for example, clumps have been observed in the accretion flow onto the UC H II region G10.6–0.4 [108, 188]). Our data do not allow us to distinguish between these two possibilities.

The additional mass required to increase the density can be estimated. The ionized radius r_S scales roughly with the density n as $r_s \propto n^{-2/3}$ (if the radius of the H II region is set by a balance between photoionization and recombination) and at optically thick frequencies the flux density scales as $S_v \propto r_s^2$. The ionized mass within G24 A1 is $M_{HII} \sim 1 \times 10^{-2} M_{\odot}$ (Sect. 4.4.1). Therefore, the sudden accretion of $\sim 5 \times 10^{-3} M_{\odot}$ into the H II region would suffice to explain the observed 45% flux decrease. The molecular accretion flow around this H II region has a mass of $\sim 130 M_{\odot}$ [20]. Thus, the required variation, 50% of the mass of the H II region, is only of the order of 0.001% of the total mass of the accretion flow.

Finally, we shall mention that even when the observations here presented are not consistent with a simple, continuous expansion of G24 A1, our result does not mean that this H II region does not expand at all, or that it is contracting continuously. As will be shown in the following chapter, an emerging model of the early evolution of H II regions embedded in massive accretion flows indicates that H II region variability is a common signature of the interaction of the neutral and ionized portions of the accretion flow in which clusters of massive stars form.

4.5 Conclusions

Our analysis of archival VLA observations of the HC H II region G24.78 + 0.08 A1 indicates a contraction of its radius between 1984.36 and 1989.48. This finding is consistent with the hypothesis that this HC H II is the inner, ionized part of the larger scale accretion flow seen in the molecular line observations of Beltrán et al. [20, 22].

Chapter 5
Time Variability of HII Regions in Numerical Simulations of MSFR

5.1 Summary

Ultracompact and hypercompact H II regions appear when a star with a mass larger than about 15 solar masses starts to ionize its own environment. Recent observations of their time variability, like the one presented in the previous chapter, are one of the pieces of evidence that suggest that at least some of them harbor stars that are still accreting from an infalling neutral accretion flow that becomes ionized in its innermost part. We present an analysis of the properties of the H II regions that are formed in the 3D radiation-hydrodynamic simulations presented by Peters et al. [161] as a function of time. Flickering of the H II regions is a natural outcome of this model. The radio-continuum fluxes of the simulated H II regions, as well as their flux and size variations are in agreement with the available observations. From the simulations, we estimate that a small but non-negligible fraction ($\sim$10%) of observed H II regions should have detectable flux variations (larger than 10%) on timescales of $\sim$10 years, with positive variations being more likely to happen than negative variations. A novel result of these simulations is that negative flux changes do happen, in contrast to the simple expectation of ever growing H II regions. We also explore the temporal correlations between properties that are directly observed (flux and size) and other quantities like density and ionization rates.

These results have been published in Galván-Madrid, Roberto, Peters, Thomas, Keto, Eric, Mac Low, Mordecai-Mark, Banerjee, Robi, and Klessen, Ralf S. "Time variability in simulated ultracompact and hypercompact HII regions." Monthly Notices of the Royal Astronomical Society, 416, 1033 (2011) [76].

R.J. Galván-Madrid, *On the Formation of the Most Massive Stars in the Galaxy*,
Springer Theses, DOI 10.1007/978-1-4614-3308-8_5,
© Springer Science+Business Media New York 2012

5.2   Introduction

The most massive stars in the Galaxy, O-type stars with masses $M_\star > 20\,M_\odot$, emit copious amounts of UV photons [203] that ionize part of the dense gas from which they form. The resulting H II regions are visible via their free-free continuum and recombination line radiation [146, 221]. H II regions span orders of magnitude in size, from giant ($D \sim 100\,\mathrm{pc}$) bubbles, to "ultracompact" (UC) and "hypercompact" (HC) H II regions, loosely defined as those with sizes of ~ 0.1 and $\sim 0.01\,\mathrm{pc}$ (or less), respectively (see the reviews by Churchwell [46], Kurtz [124], and Hoare et al. [90]). UC and HC H II regions are the most deeply embedded, and so are best observed at radio wavelengths.

Large, rarefied H II regions expand without interruption within the surrounding medium due to the high pressure contrast between the ionized and neutral phases [193]. This simple model was extrapolated to the ever smaller objects recognized later and is widely used to interpret observations of UC and HC H II regions. Common assumptions about these objects are: (1) They are steadily expanding within their surrounding medium at the sound speed of the ionized gas, ~ 10 km s^{-1}. (2) The ionizing star(s) is already formed, i.e., accretion to the massive star(s) powering the H II region has stopped.

However, evidence has accumulated that suggests a revision of these assumptions:

1. The hot molecular cores embedding UC and HC H II regions are often rotating and infalling [22, 24, 42, 103, 117], sometimes from parsec scales all the way to the immediate surroundings of the ionized region [17, 74].
2. Infall of gas at velocities of a few km s^{-1} directly toward the ionized center has also been observed in UC and HC H II regions [22, 74, 230].
3. The inner ionized gas has been resolved in a few cases, and it also shows accretion dynamics (outflow, infall, and rotation [74, 108, 181]).
4. The spectral index α (where the flux goes as $S_\nu \propto \nu^\alpha$) of some UC and HC H II regions is ~ 1 from cm to mm wavelengths, indicating density gradients and/or clumpiness inside the ionized gas [14, 66, 94, 114].
5. A few UC and HC H II regions have been shown to have variations on timescales of years [2, 67, 73, 206]. These variations indicate that UC and HC H II regions sometimes expand [2], and sometimes contract [73]. Some other ionized regions around massive protostars have been shown to remain approximately constant in flux [82].

All these observations strongly suggest that UC and HC H II regions *are not* homogeneous spheres of gas freely expanding into a quiescent medium, but rather that these small H II regions are intimately related to the accretion processes forming the massive stars. Simple analytic models show that the observed H II regions can be either the ionized, inner part of the inflowing accretion flow [104, 105] or the ionized photoevaporative outflow [93] fed by accretion [106].

Numerical simulations of the formation and expansion of H II regions in accretion flows around massive stars have only recently become possible with three-dimensional radiation-hydrodynamics. Studies that simulate the expansion of H II regions have focussed on larger-scale effects on the parental molecular cloud [12, 50–52, 84, 160]. However, in none of those simulations was the ionizing radiation produced by the massive stars dynamically forming through gravitational collapse in the molecular cloud. Recent simulations by Peters et al. [161] include a more realistic treatment of the formation of the star cluster. Radio-continuum images generated from the output of those simulations show time variations in the morphology and flux from the H II regions produced by the massive stars in formation. These changes are the result of the complex interaction of the massive filaments of neutral gas infalling to the central stars with the ionized regions produced by some of them. A statistical analysis of the H II region morphologies [162] consistently reproduces the relatively high fraction of spherical and unresolved regions found in observational surveys. Thus, the non-monotonic expansion of H II regions, or flickering, appears able to resolve the excess number of observed UC and HC H II regions with respect to the expectation if they expand uninterrupted (the so-called lifetime problem put forward by Wood and Churchwell [221]). Furthermore, the ionizing radiation is unable to stop protostellar growth when accretion is strong enough. Instead, accretion is stopped by the fragmentation of the gravitationally unstable accretion flow in a process called "fragmentation-induced starvation", a theoretical discussion of which can be found in Peters et al. [163].

In this chapter we present a more detailed analysis of the flux variability in the simulations presented in Peters et al. [161]. In Sect. 5.2 of this chapter we describe the set of numerical simulations and the methods of analysis. In Sect. 5.3 we present our results. Section 5.4 discusses the implications of our findings. In Sect. 5.5 we present our conclusions.

5.3 Methods

5.3.1 The Numerical Simulations

Our study uses the highest resolution simulations of those presented in Peters et al. [161]. The simulations use a modified version of the adaptive-mesh code FLASH [69], including self-gravity and radiation feedback. They include for the first time a self-consistent treatment of gas heating by both ionizing and non-ionizing radiation. Peters et al. [161] describes in detail the numerical methods. The initial conditions are a cloud mass of $1,000 M_\odot$ with an initial temperature of $30 \, \mathrm{K}$. The initial density distribution is a flat inner region of $0.5 \, \mathrm{pc}$ radius surrounded by a region with a decreasing density $\propto r^{-1.5}$. The density in the homogeneous volume is $1.27 \times 10^{-20} \, \mathrm{g} \, \mathrm{cm}^{-3}$. The simulation box has a length of $3.89 \, \mathrm{pc}$. The size and mass of the cloud are in agreement with those of star-formation regions that are able to produce at least one star with $M_\star > 20\,M_\odot$ [74].

Run A (as labeled in Peters et al. [161]) has a maximum cell resolution of 98 AU and only the first collapsed object is followed as a sink particle [63]. In this run, the formation of additional stars (the sink particles) is suppressed using a density-dependent temperature floor (see Peters et al. [161] for details). On the other hand, in Run B additional collapse events are permitted and a star cluster is formed, each star being represented by a sink particle. The maximum resolution in Run B is also 98 AU.

5.3.2   Data Sets

For the entire time span of Run A (single sink) and Run B (multiple sink), radio-continuum maps at a wavelength of 2 cm were generated from the simulation output every $\sim$300 year by integrating the radiative transfer equation for free-free radiation while neglecting scattering [83]. Following Mac Low et al. [139], each intensity map was then convolved to a circular Gaussian beam with half-power beam width HPBW = 0.14" (assuming a source distance of 2.65 kpc). A noise level of 10^{-3} Jy was added to each image. Further details are given in Peters et al. [161]. These maps were used to explore the behavior of the free-free continuum from the H II region over the entire time evolution of the simulations. For Run B, sometimes the H II regions overlap both physically in space and/or in appearance in the line of sight. Therefore, the presented time analysis refers to the entire star cluster unless otherwise specified.

To compare more directly to available observations, which span at most a couple of decades in time, each of Run A and B were re-run in four time intervals for which a flux change was observed in the radio-continuum images mentioned above. Data dumps and radio-continuum maps were generated at every simulation time step ($\sim$10 year). The analysis performed in the low time-resolution maps was also done in these high time-resolution data. For Run B, the intervals for the high time-resolution data were also selected such that the H II region powered by the most massive star is reasonably isolated from fainter H II regions ionized by neighboring sink particles, both in real space and in the synthetic maps.

5.4   Results

5.4.1   Variable HII Regions

Real UC and HC H II regions, as well as those that result from the simulations presented here, are far from the ideal cases described by, e.g., Spitzer [193].

However, it is instructive to discuss the limiting ideal cases to show that their variability is a natural consequence of their large to moderate optical depth.

The flux density S_v of an ionization-bounded[1] H II region is

$$S = \frac{2kv^2}{c^2} \int_\Omega T_B d\Omega, \tag{5.1}$$

where k is the Boltzmann constant, c is the speed of light, v is the frequency, and the brightness temperature T_B is integrated over the angular area Ω of the H II region.

In the limit of very low free–free optical depths ($\tau_{ff} \ll 1$), T_B along a line of sight l goes as

$$T_B \propto T_e^{-0.35} v^{-2.1} \int_l n^2 dl, \tag{5.2}$$

where T_e and n are the electron temperature and density respectively.

Combining Equations (5.1) and (5.2) we have that for a given v and constant T_e:

$$S(\tau_{ff} \ll 1) \propto n^2 R^3 \propto \dot{N}, \tag{5.3}$$

where the last proportionality comes from the Strömgren relation $\dot{N} \propto n^2 R^3$ ($\dot{N}$ is the ionizing-photon rate and R is the 'radius' of the H II region). Equation 5.3 shows that in the optically-thin limit the H II -region flux only depends on $\dot{N}$. For time intervals of a few $\times 10$ year the mass and ionizing flux of an accreting protostar remain almost constant, and so does the flux of an associated optically-thin H II region.

On the other hand, for very high optical depths ($\tau_{ff} \gg 1$) $T_B = T_e$. For a given frequency and constant T_e, (5.1) becomes:

$$S(\tau_{ff} \gg 1) \propto R^2 \propto \dot{N}^{2/3} n^{-4/3}, \tag{5.4}$$

therefore, the flux of an optically thick H II region is proportional to its area, and both flux and area decrease with density.

This analysis is valid for time intervals larger than the recombination timescale (~ 1 month for $n \sim 10^6 \, cm^{-3}$, see e.g., Osterbrock [157]) and as long as the growth of $\dot{N}$ is negligible.

However, the H II regions in the simulations are clumpy and have subregions of high and low free–free optical depth. Their flux during the accretion stage (while they flicker) is dominated by the denser, optically thicker ($\tau_{ff} > 1$) subregions, so their behaviour is closer to (5.4) than to (5.3). The on-line version of this paper contains a movie of τ_{ff} for Run B as viewed from the Z-axis (line of sight perpendicular to the plane of the accretion flow). The clumpiness and intermediate-to-large optical depth of these H II regions are also the reasons behind their

---

[1] For H II regions that are embedded in their parental cloud, the ionization-bounded approximation is better than the density-bounded approximation. The analysis here presented can be derived from, e.g., Mezger and Henderson [146], Spitzer [193], Rybicki and Lightman [178], and Keto [105].

rising spectral indices up to relatively large frequencies ($v > 100\,\mathrm{GHz}$) without a significant contribution from dust emission (see the analytical discussions of Ignace and Churchwell [94] and Keto et al. [114], or the analysis of these simulations in Peters et al. [162]). As for the variability, the large optical depths cause the size and flux of the simulated H II regions to be well correlated with each other, and anticorrelated with the density of the central ionized gas (see Sect. 3.6). The neutral accretion flow in which the ionizing sources are embedded is filamentary and prone to gravitational instability (further discussion is in Peters et al. [162, 163]). The changes in the density of the H II regions are a consequence of their passage through density enhancements in the quickly evolving accretion flow.

5.4.2   Global Temporal Evolution

Figure 5.1 shows the global temporal evolution of the H II region in Run A (single sink particle). The 2-cm flux ($S_{2\mathrm{cm}}$) observed from orthogonal directions and the mass of the ionizing star ($M_\star$) are plotted against time. The global temporal trend of the H II region is to expand and become brighter. However, fast temporal variations are seen at all the stages of the evolution. The fluxes in the projections along the three different cartesian axes follow each other closely. For the rest of the analysis, the Z-axis projection, a line of sight perpendicular to the plane of the accretion flow, is used.

The H II region is always faint ($S_{2\mathrm{cm}} < 1\,\mathrm{Jy}$ at the assumed distance of 2.65 kpc) for $M_\star < 25\,M_\odot$. Past this point, the H II region is brighter than 1 Jy 86% of the time (Fig. 5.1).

A similar analysis of the H II region around the most massive star in Run B (multiple sinks) is shown in Fig. 5.2. Only the Z-axis projection, i.e., a line of sight perpendicular to plane of the accretion flow is used, since only from this viewing angle the brightest H II region is well separated from other H II regions at all times (the flux movies are presented in Peters et al. [161]). The extra fragmentation in Run B translates into a weaker accretion flow and a lower-mass ionizing star as compared to that of Run A, a process referred to as "fragmentation-induced starvation" (a theoretical discussion of this process in presented in Peters et al. [163]). Therefore, the brightest H II region in Run B (Fig. 5.2) is weaker than the H II region in Run A (Fig. 5.1). Figures 5.1 and 5.2 do not show times later than $t = 100\,\mathrm{kyr}$ in order to facilitate their comparison. Both runs continue past this time, but in Run B accretion onto the most massive star stops at $t = 109\,\mathrm{kyr}$, while in Run A the artificial suppression of the fragmentation leads to an unrealistically large mass for the ionizing star at later times.

H II regions are highly variable both in Run A and Run B. However, since the suppression of fragmentation in Run A produces a larger accretion flow and a most massive ionizing star, Run A presents larger flux variations than Run B. Figure 5.3 shows the flux changes over the evolution of the H II regions in both runs. A comparison of the fractional variations of the H II regions shows that, though

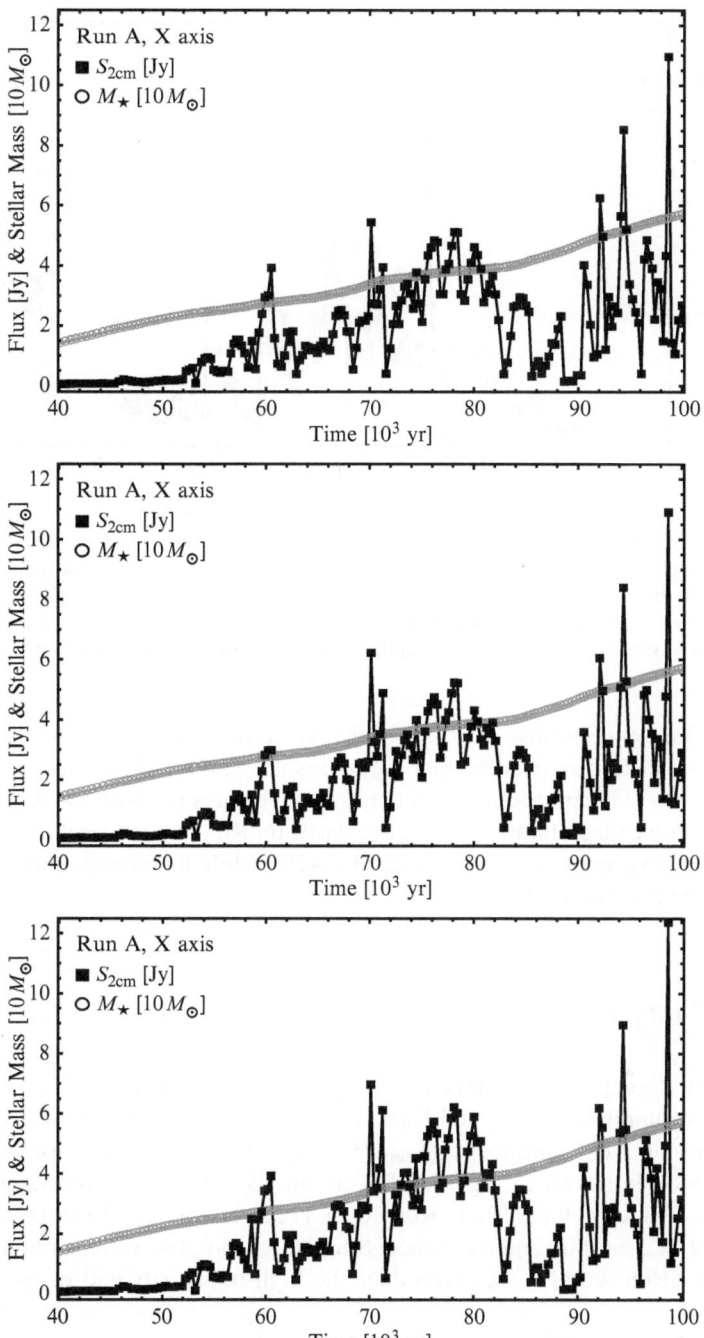

Fig. 5.1   2-cm flux ($S_{2\,cm}$, filled *black squares*) and stellar mass ($M_\star$, *gray circles*) as a function of time (t) for Run A. Although the long-term trend of the H II region is to increase in flux, it constantly flickers during its evolution. The fluxes at the different projections (X-axis *top*, Y-axis *middle*, Z-axis *bottom*), though not the same, follow each other closely

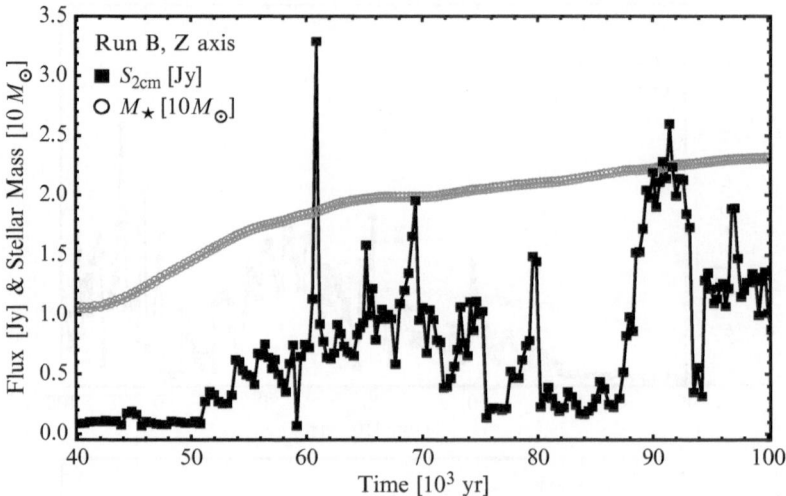

Fig. 5.2 2-cm flux (S_{2cm}, filled *black squares*) of the H II region formed by the most massive star and its stellar mass ($M_\star$, *gray circles*) as a function of time (t) for Run B. Although the long-term trend of the H II region is to increase in flux, it constantly flickers during its evolution. Only the Z-axis projection is used because from this viewing angle, perpendicular to the plane of the accretion flow, the brightest H II region can be distinguished from other H II regions at all times

more similar between runs, they are still larger in Run A (Fig. 5.4). For consecutive data points, in Run A positive flux variations are 56% of the events and the flux increment is +74% on average, while negative changes are 44% of the events and have an average magnitude of −27%. Similarly, for Run B, positive changes (52% of the events) have an average magnitude of +42%, while the average flux decrement (48% of the events) is −18%.

5.4.3 Comparison to Surveys

We present a comparison with the ionized-gas surveys of UC and HC H II regions by Wood and Churchwell [221] and Kurtz et al. [125].[2] Figure 5.5 shows normalized histograms of 2-cm luminosity ($S_{2cm}d^2$) obtained for both runs using the time steps previously shown in Figs. 5.1–5.4 and the co-added observed samples from Wood and Churchwell [221] and Kurtz et al. [125], taking the 81 sources for which they report a 2-cm flux and a distance. Simulation and observations roughly agree, but neither Run A nor B can reproduce the high-luminosity end of the observed

---

[2]The relation of the surrounding molecular gas to the ionized gas has not been explored in detail for many of their sources, which makes difficult to assess whether they are candidates to harbor accreting protostars.

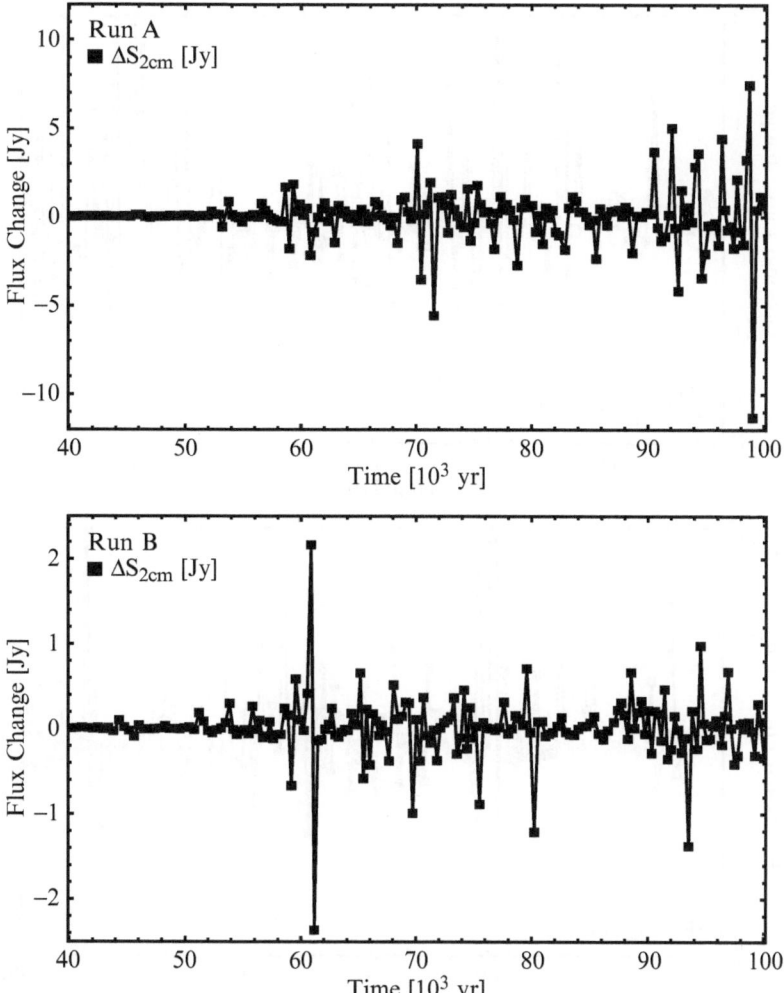

Fig. 5.3 2- cm flux variations (ΔS_{2cm}) for Run A (*top* panel) and Run B (*bottom* panel)

distribution: 20% of the observed sources have $S_{2cm}d^2 > 50$ Jy kpc^2, while only 1% of the H II regions in the simulation steps of Run A and in no step in Run B have luminosities above this threshold. These bright UC H II regions likely correspond to stages in which accretion to the ionizing protostar(s) is completely shut off, therefore they do not correspond to the analyzed H II regions from the simulation in which the protostars are still accreting. In the range $20 < S_{2cm}d^2 < 50$ Jy kpc^2, Run A matches better than Run B the observed luminosities. This may be because Run B does not produce any star with a mass $M_\star > 30 M_\odot$. Simulations identical to Run B produce higher-mass stars in the presence of magnetic fields [164] and may also produce more massive stars in the purely radiation-hydrodynamical case with higher-mass

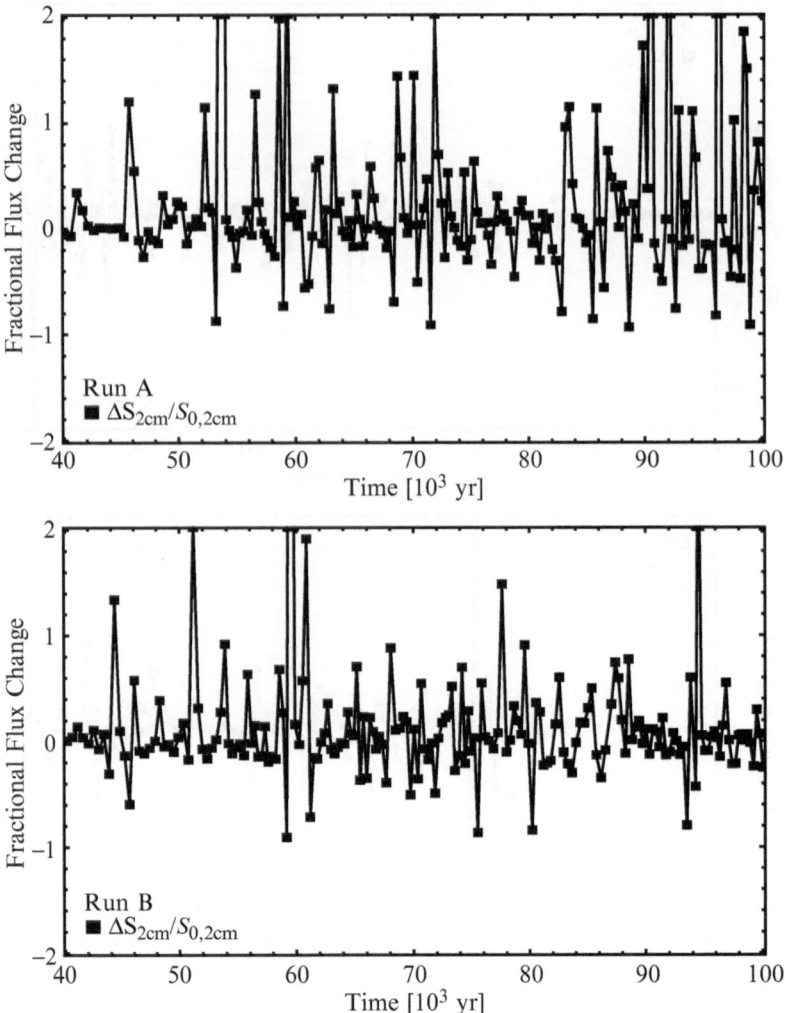

Fig. 5.4 Fractional flux variations ($\Delta S_{2\text{cm}}/S_{2\text{cm},0}$) for Run A (*top panel*) and Run B (*bottom panel*)

initial clumps. We plan to perform in the future a study on the robustness of our results for different initial conditions. For the smallest luminosities, Run B matches better than Run A the observed 55% of H II regions that have $S_{2\text{cm}}d^2 < 5\,\text{Jy}$ kpc^2, but over-estimates (35%) the observed fraction (14%) of H II regions with $5 < S_{2\text{cm}}d^2 < 10$. Peters et al. [162] presented statistics of the morphologies of the H II regions in Run A and Run B and found that Run B agrees better with observed surveys. Although Run A can be interpreted as a mode of isolated massive star formation, the treatment of fragmentation is more realistic in Run B. Moreover, most high-mass stars form in clusters [238].

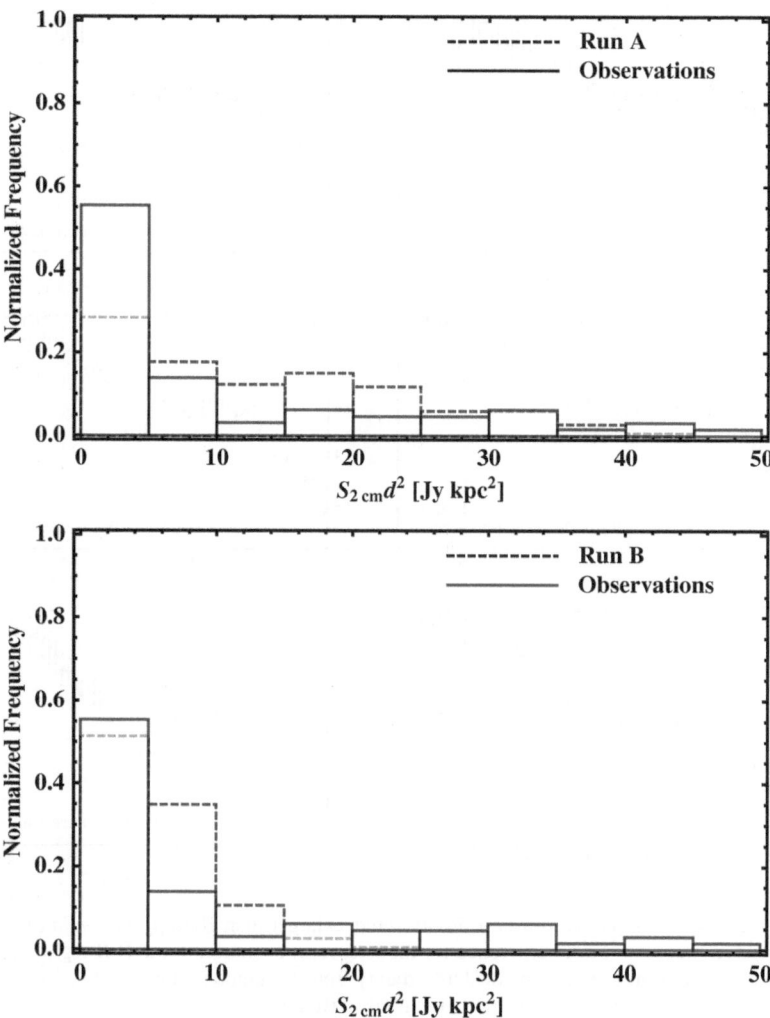

Fig. 5.5 Histograms of 2-cm luminosities ($S_{2cm}d^2$) for the global temporal evolution of the simulated H II regions (*dashed lines*) and the co-added samples of Wood and Churchwell [221] and Kurtz et al. [125] (*solid lines*). The *top* and *bottom* frames correspond to Run A and Run B respectively

5.4.4 Long-Term Variation Probabilities

We address the question of the flux variability expected from the simulations by calculating the probability of variations larger than a given threshold as a function of time difference between steps. The low time-resolution data has the advantage of spanning the entire runs, but is not useful to predict the expected variations on timescales shorter than 10^2 year. We use the high time-resolution data sets to make

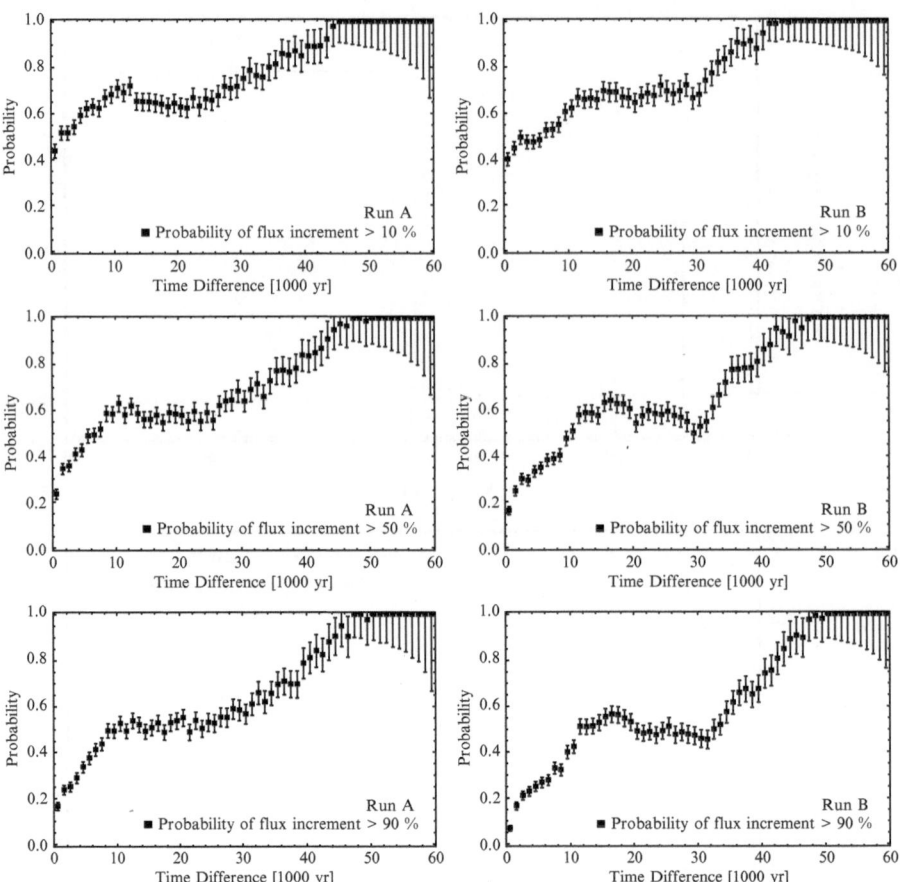

Fig. 5.6 *Left column*: probabilities for flux increments larger than 10 (*top*), 50 (*middle*), and 90% (*bottom*) as a function of time difference for the long-term evolution of the H II region in Run A. The error bars indicate the 1σ statistical uncertainty from the number of counts in each bin 1,000-year wide. *Right column*: same for the H II regions in Run B

an estimate of the flux variations over shorter timescales, but we caution that the analyzed time intervals may not be representative of the entire simulation. We use this approach because re-running the entire simulations to produce data at ∼10 year resolution is not feasible.

Figure 5.6 shows the probabilities of *flux increments larger than a given threshold* for time differences between 1 and 60 kyr for Run A (left column)and Run B (right). The *top* panels correspond to flux increments larger than 10%, the *middle* panels correspond to 50%, and the *bottom* panels to 90%. On average, H II regions tend to expand, making a given flux increment to be more likely to happen for larger time intervals than for shorter ones.

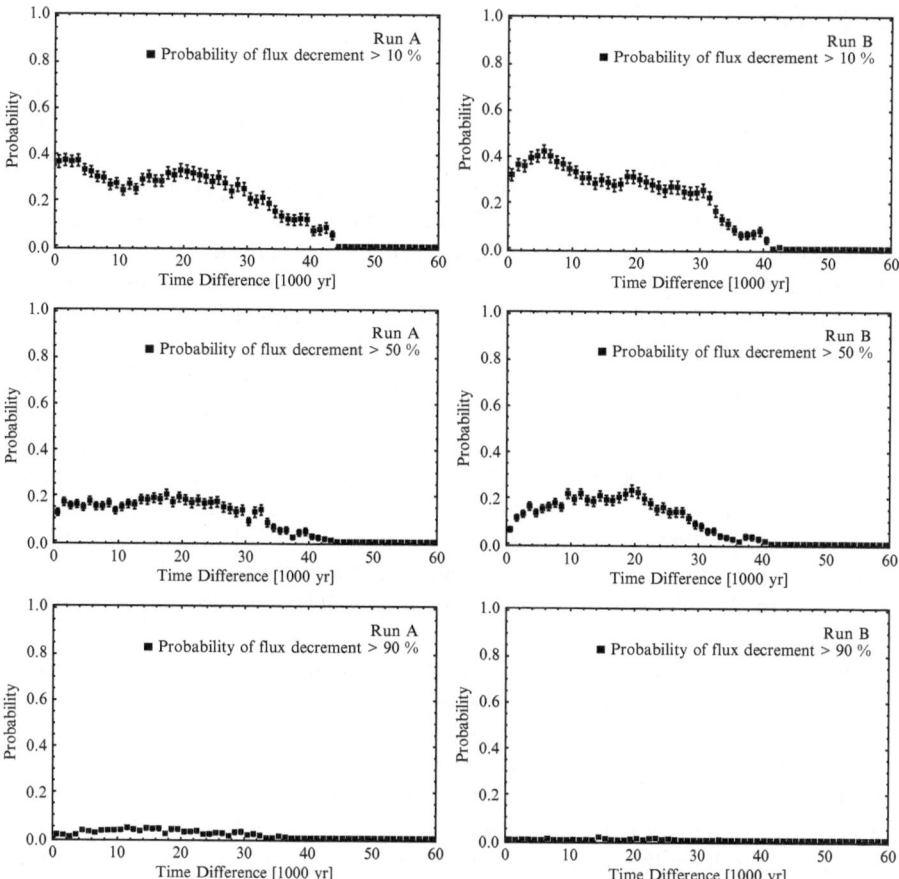

Fig. 5.7 *Left column*: probabilities for flux decrements larger than 10 (*top*), 50 (*middle*), and 90% (*bottom*) as a function of time difference for the long-term evolution of the H II region in Run A. The error bars indicate the 1σ statistical uncertainty from the number of counts in each bin 1,000-year wide. *Right column*: same for the H II regions in Run B

Figure 5.7 show the probabilities of *flux decrements larger than a given threshold* (in modulus) for Run A (left column) and Run B (right) respectively. The top to bottom order of the thresholds is as in Fig. 5.6. At $\Delta t > 30$ kyr the negative-change probabilities decrease and reach ~ 0 at about $\Delta t > 40$ kyr for any threshold. This is caused by the eventual growth of the H II regions in spite of the flickering. There is some indication that the probabilities for negative changes also decrease at timescales shorter than 1 kyr, specially for flux-change thresholds larger than 50% (see Fig. 5.7). This is also suggested from the analysis of the high temporal-resolution data in the next section. Although flux increments are more likely than decrements for any given threshold and time lag, a novel result of these simulations is that negative flux changes do happen, in contrast with the simple expectation of ever growing H II regions.

5.4.5 Short-Term Variation Probabilities

We re-ran four time intervals in each of Run A and Run B, spanning a few hundred years each, and producing data dumps at each simulation step ($\sim$10 year). These time intervals were selected to contain a pair of negative/positive flux changes to investigate the correlation of flux variations with physical changes in the H II region, like size and density (next section). Therefore, they may not be representative of the entire simulation, but since it is not feasible to re-run the entire simulations producing data dumps at the highest temporal resolution, we use these data to constrain the expected flux variations in observable timescales. We argue that the close match at scales of $\Delta t \sim 500$ year between the probabilities obtained from the low temporal-resolution (previous section) and the high temporal-resolution data (this section) indicates that the results presented here are meaningful.

Figure 5.8 shows the probabilities for flux increments larger than the specified threshold as a function of time lag. This figure is the analog of Fig. 5.6 for the high temporal-resolution data. The slight probability decrements after $\Delta t \sim 300$ year, especially at low thresholds, are an artifact caused by the fact that the data sets include a negative/positive flux-change pair. Still, the probabilities at $\Delta t = 490$ year match within 20–80% with the probabilities at $\Delta t = 500$ year from the low-temporal resolution data.

On observable timescales, $\Delta t = 0$ to 40 year, a small but non-zero fraction of H II regions is expected to have detectable flux increments. For two observations separated by 10 year, Run A gives a prediction of $16.7 \pm 2.9\%$ of H II regions having flux increments larger than 10%, $6.8 \pm 1.9\%$ with flux increments larger than 50%, and $4.7 \pm 1.6\%$ with increments larger than 90%. The more realistic Run B predicts a smaller fraction of variable H II regions: $6.9 \pm 1.6\%$, $0.3 \pm 0.3\%$, and 0% of them are expected to have flux increments larger than 10%, 50%, and 90% over a time interval of 10 year, respectively.

Figure 5.9 show the probabilities for flux decrements larger than a given threshold obtained with the high temporal-resolution data. The probabilities obtained at $\Delta t = 490$ year from the high temporal-resolution data roughly match with those at $\Delta t = 500$ year obtained from the low temporal-resolution data, within a factor of 1–3.

Negative variations should also be detectable in a non-negligible fraction of H II regions. Run A predicts that $5.7 \pm 1.7\%$, $2.6 \pm 1.2\%$, and $1.6 \pm 0.1\%$ of H II regions should present flux decrements larger than 10%, 50%, and 90% respectively, when observed in two epochs separated by 10 year. Run B predicts a smaller fraction of H II regions with negative flux variations: $3.3 \pm 1.1\%$, $1.5 \pm 0.7\%$, and 0% for thresholds at 10%, 50%, and 90% respectively. We emphasize that negative variations are not expected in classical models of monotonic H II region growth, while they are a natural outcome of the model presented here. Moreover, these variations should be detectable with current telescopes.

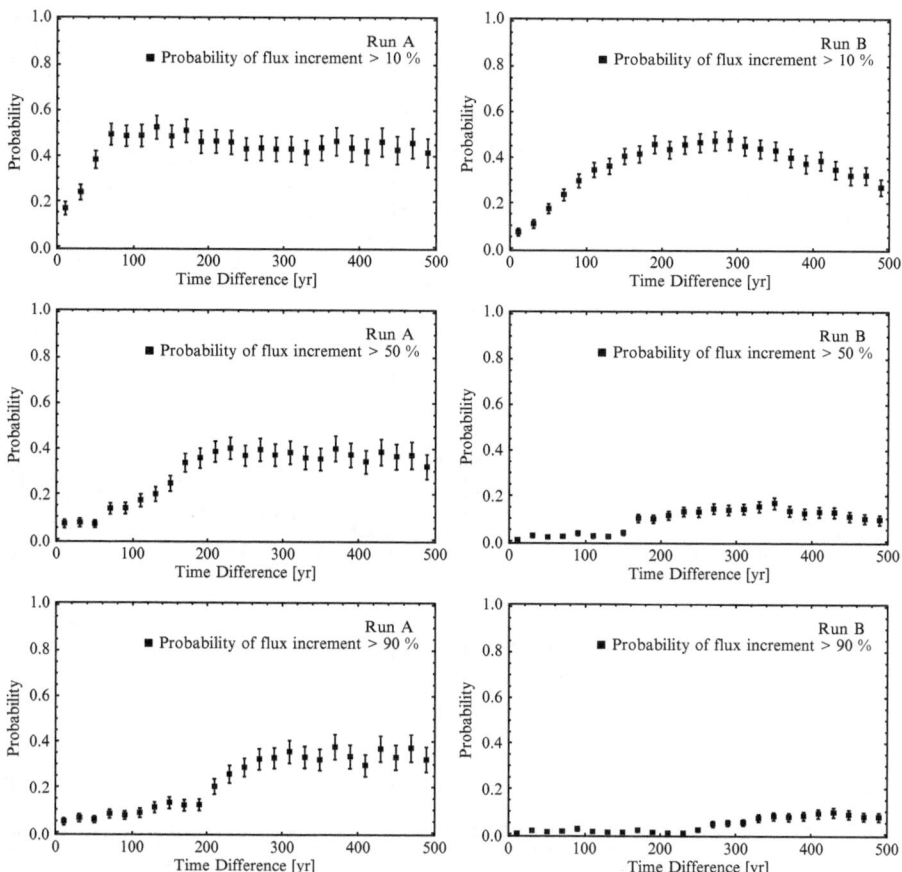

Fig. 5.8 *Left column*: probabilities for flux increments larger than 10 (*top*), 50 (*middle*), and 90% (*bottom*) as a function of time difference for the sample intervals at high time resolution. The error bars indicate the 1σ statistical uncertainty from the number of counts in each bin 20-year wide. *Right column*: same for the H II region around the most massive star in Run B

5.4.6 Variations in Other Properties of the HII Regions

Because Run B is more realistic in the treatment of fragmentation (see Peters et al. [161]), we use the time intervals of Run B with data at high temporal resolution to investigate the correlations of sudden flux changes with other properties of the H II regions.

Let the scale length L_{HII} of the H II region of interest be defined as

$$L_{\mathrm{HII}} = 2(A/\pi)^{1/2}, \qquad (5.5)$$

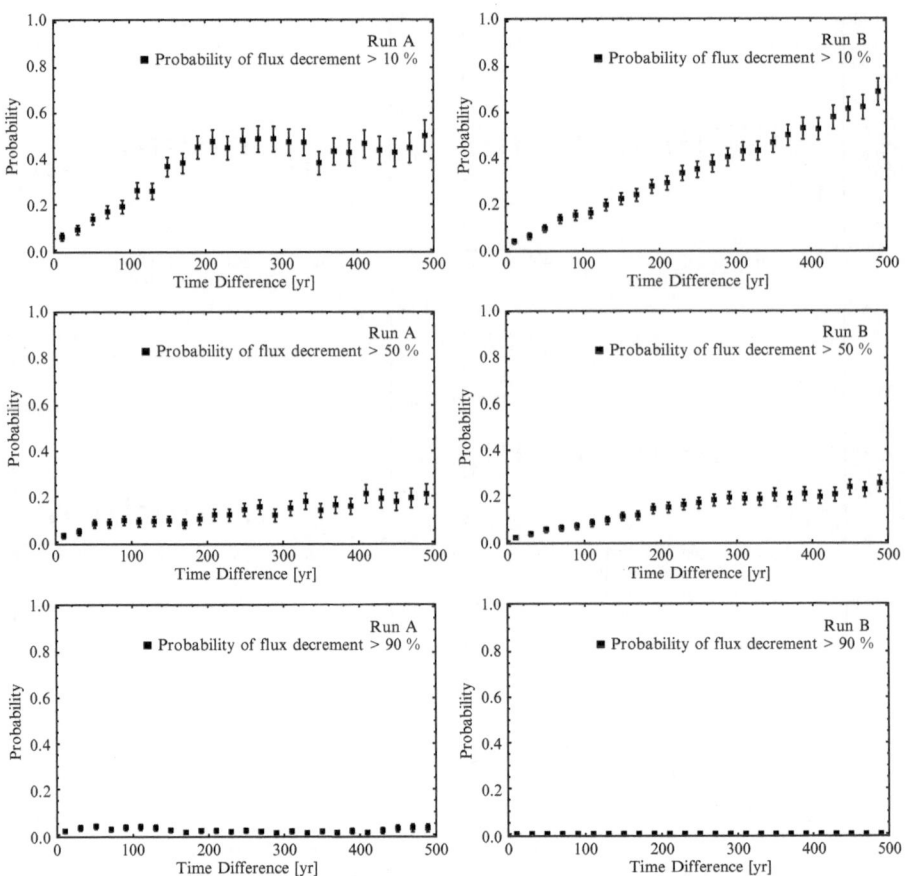

Fig. 5.9 *Left column*: probabilities for flux decrements larger than 10 (*top*), 50 (*middle*), and 90% (*bottom*) as a function of time difference for the sample intervals at high time resolution. The error bars indicate the 1σ statistical uncertainty from the number of counts in each bin 20-year wide. *Right column*: same for the H II region around the most massive star in Run B

where A is the area in the synthetic image where the H II region is brighter than three times the rms noise. Figure 5.10 compares the time evolution of L_{HII}, the ionized-gas density ρ_{HII} within the H II region, and the rate of ionization of neutral gas $\dot{M}_{\rightarrow HII}$.

Figure 5.11 further compares the time evolution of the 2-cm flux S_{2cm}, the total ionized-gas mass M_{HII} in the H II region, and the ionized-gas mass within the same volume with density $\rho_{HII} > 10^{-17}$ g cm^{-3}, the typical peak density reached when the H II region gets denser immediately after the flickering events (see Fig. 5.10).

The flux-size correlation, as well as the size-density anticorrelation are a consequence of the relatively large optical depths of the H II regions (see Sect. 3.1). The "quenching" events, when an H II region has a large, sudden drop in flux and size, are coincident with large increments in the ionized density. The H II region

Fig. 5.10 Variation at high temporal resolution in Run B of the scale length L_{HII} (*filled black circles*) of the H II region around the most massive sink particle, its density ρ_{HII} (*filled black squares*), and its rate of ionization of neutral gas $\dot{M}_{\rightarrow HII}$ (*filled black diamonds*)

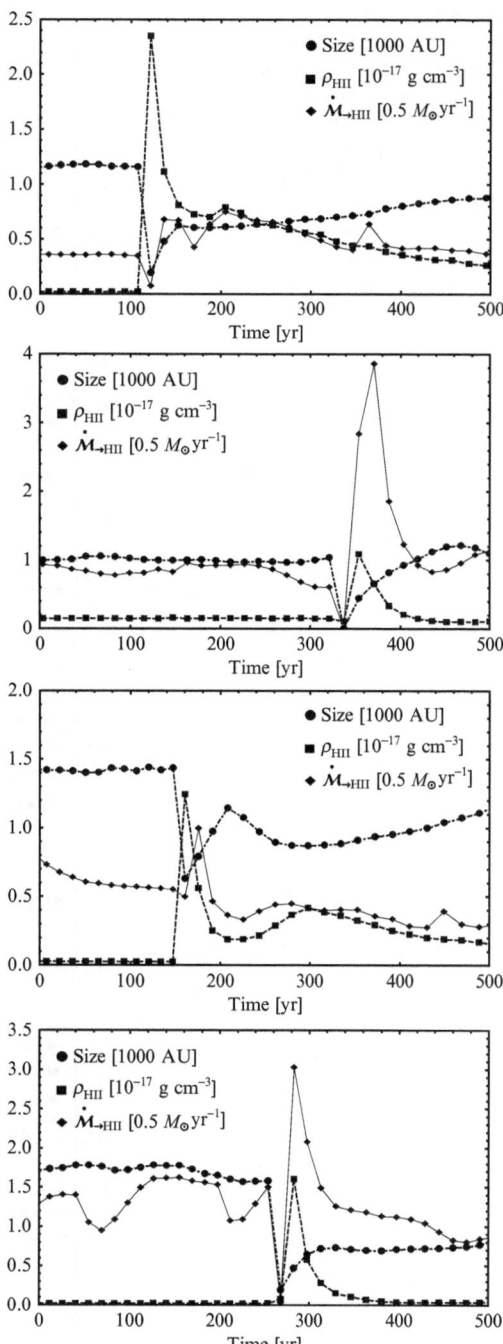

Fig. 5.11 Variation at high temporal resolution in Run B of the 2-cm flux $S_{2\,cm}$ (*filled black circles*) of the H II region around the most massive sink particle, its ionized mass M_{HII} (*filled black squares*), and its dense-gas ionized mass $M_{HII,dense}$ (*filled black diamonds*)

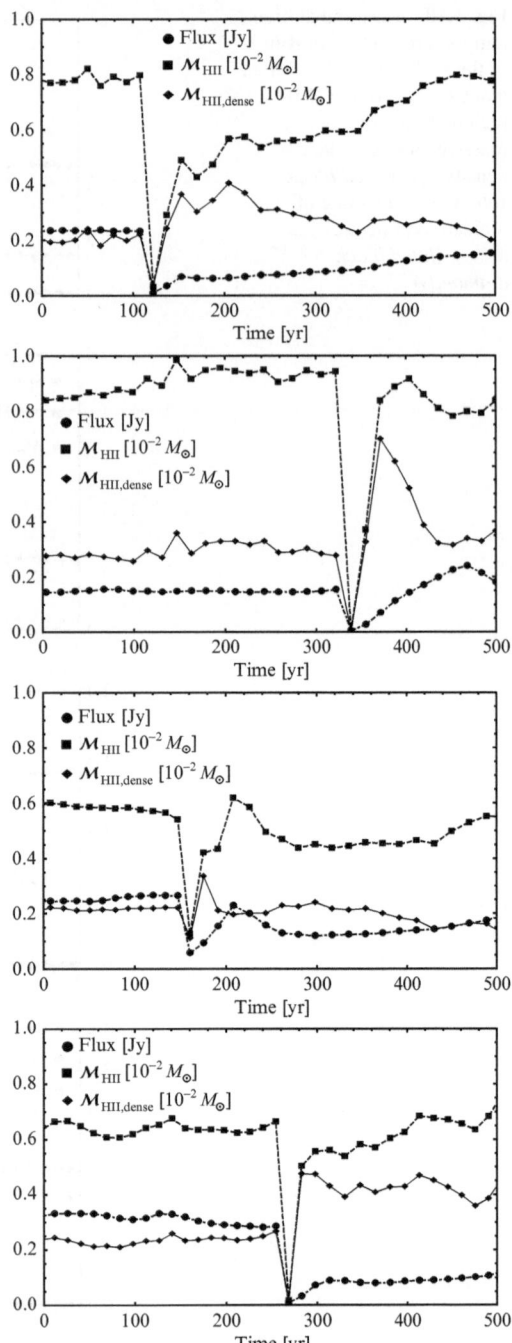

rapidly reaches a state close to ionization equilibrium at its new size after the quenching instability. In Fig. 5.10 it is seen that at the moment of the quenching, the ionization rate has a sharp decrement immediately followed by an even faster increment that marks the initial re-growth of the H II region. Shortly afterwards, the ionization rate stabilizes again and the H II region grows hydrodynamically, gradually becoming larger and less dense.

The ionized mass of the H II regions follows the flux and size closely (Fig. 5.11), since this is the gas responsible for the free-free emission. However, if only the denser gas is taken into account, the amount of this dense ionized gas is particularly high in the initial re-growth of the H II region immediately after the quenching event (Fig. 5.11). Therefore the rapid quenching events are marked by the presence of denser gas around the accreting massive protostar.

5.5 Discussion

5.5.1 A New View of Early H II Region Evolution

Until recently, H II regions have been modeled as (often spherical) bubbles of ionized gas freely expanding into a quiescent medium. However, this paradigm fails to explain observations of some UC and HC H II regions (see Sect. 5.1). There is enough evidence to assert that massive stars form in clusters by accretion of gas from their complex environment (reviews have been presented by Garay and Lizano [77], Mac Low and Klessen [138], Beuther et al. [30], and Zinnecker and Yorke [238]). Centrally peaked, anisotropic density gradients are expected at the moment an accreting massive star starts to ionize its environment. Therefore, the earliest H II regions should not be expected to be the aforementioned bubbles, but more complex systems where ionized gas that is outflowing, rotating, or even infalling may coexist. The feasibility of this scenario has been shown by analytic models [93, 104–106, 137] and recent numerical simulations [161–163]. In these numerical models, the inner part of the accretion flow is mostly ionized, while the outer part is mostly neutral. The neutral accretion flow continuously tries to feed the central stars. The interaction of this infalling neutral gas with the ionized region has a remarkable observational effect: the flickering of the free–free emission from the H II region.

We stress that the flickering is not a gasdynamical effect. Instead, it is a non-local result of the shielding of the ionizing source by its own accretion flow. Hence, variations in the ionization state of the gas inside the H II region are not limited by the speed of sound but can happen on the much shorter recombination timescale of the ionized gas, rendering direct observation of this flickering effect feasible.

5.5.2   Observational Signatures

Time-variation effects are expected in the ionized gas but not in the molecular gas, since small clumps of molecular gas that become ionized and/or recombine have a much larger effect on the emission of the $< 1\, M_\odot$ of ionized gas than on the tens to hundreds of $M_\odot$ of molecular gas in the accretion flow. One observational example is the HC H II region G20.08 N A, for which Galván-Madrid et al. [74] reports an ionized mass of $0.05\, M_\odot$ and a mass of warm molecular gas in the inner 0.1 pc of 35 to $95\, M_\odot$ (another well studied example is G10.6–0.4, with more than $1,000\, M_\odot$ in the pc-scale molecular flow surrounding the H II regions [103, 108, 133]). The H II regions in the simulations also have masses of ionized gas in the range 10^{-3}– $10^{-1}\, M_\odot$ while the stars are still accreting.

In Sects. 3.4 and 3.5 we have attempted to quantify the expected flux variations in the radio continuum of H II regions for massive stars forming in isolation (Run A) and, more realistically, in clusters (Run B). The accretion flow in Run A is stronger than in Run B (actually the star in Run A never stops accreting, see Peters et al. [161]), which leads to a brighter and more variable H II region.

For a given run, flux variation threshold, and time difference, positive variations are more likely to happen than negative ones, i.e., there is a constant struggle between the H II region trying to expand and the surrounding neutral gas trying to confine it, with a statistical bias toward expansion.

Monitoring H II regions for thousands of years is not possible, but a few observations of rapid flux changes over timescales of $\sim$10 year have been presented in the literature [2, 67, 73, 206]. Comparing multi-epoch images made with radio-interferometers can be challenging: even if the absolute flux scale of the standard quasars is known to better than 2%, slight differences in observational parameters between observations (mainly the Fourier-space sampling, see Perley [159]) make questionable any observed change smaller than 10%. We have therefore measured the variation probabilities in the simulations at thresholds starting at 10%. Considering this detection limit for the more realistic case of Run B, we have estimated that about 7% of UC and HC H II regions observed at two epochs separated by about 10 years should have detectable flux increments, and that about 3% should have detectable decrements. In total, $\sim$10% of H II regions should have detectable flux variations in a period of 10 years. Dedicated observations of as many sources as possible are now needed to test this model.

Our long timescale data can only be constrained by observational surveys, not by time monitoring. In Sect. 3.3 we have shown that the radio luminosities (i.e., distance-corrected flux) in the long-term evolution of the simulated H II regions are consistent with major surveys, except for the most luminous H II regions, in which likely accretion has stopped and which therefore do not correspond with our data. The hypothesis that a considerable fraction of observed UC and HC H II regions may harbor stars that are still accreting material still needs more convincing evidence in addition to matching the model here presented. For most cases the dynamics of the surrounding molecular gas and of the ionized gas have not been studied at high angular resolution, and such studies in many sources are key to test this idea. From available observations, almost all of the massive star formation regions with signatures of active accretion and in which the mass of the protostar is estimated from dynamics to be $M_\star > 20 M_\odot$ have a relatively bright H II region (with at least ~ 100 mJy at short cm wavelenghts, e.g., Beltrán et al. [23] and Galván-Madrid et al. [74]). To our knowledge, the only clear exception is the recent report by Zapata et al. [228] of an accreting protostar in W51 N with an estimated mass of $M_\star \sim 60 M_\odot$ and only 17 mJy at 7 mm. This object can be understood in the context of these simulations if it is in a quenched, faint state as currently observed.

5.5.3 Caveats and Limitations

As mentioned in Peters et al. [161–163] the simulations here presented do not include the effects of stellar winds and magnetically-driven jets originating from within 100 AU, which would produce outflows that are more powerful than the purely pressure-driven outflows that appear in Runs A and B.

The inclusion of stellar winds and jets may affect the results presented in this paper only quantitatively. Peters et al. [164] have shown that magnetically driven outflows from radii beyond 100 AU do not stop accretion and even channel more material to the central most massive protostars. The simulations of Wang et al. [212] also indicate that collimated outflows may be an important regulator of star formation by slowing the accretion rate but without impeding accretion. Observationally, molecular outflows tend to be less collimated for the more massive O-type protostars capable of producing H II regions than for B-type protostars [9]. Regarding the radio-continuum, it is unknown if the free-free emission from the photoionized H II regions produced by O-type protostars can coexist with the free–free emission from (partially) ionized, magnetically driven jets. Before the appearance of an H II region, these jets are detected in protostars less massive than $\sim 15 M_\odot$ [38], and even though their radio emission also appears to be variable (due to motions and interactions with the medium, see e.g., Curiel et al. [48]), their typical centimeter flux is ~ 1 mJy, one to two orders of magnitude fainter than the typical flux of UC and HC H II regions (except maybe for the youngest gravitationally-trapped H II regions [105]). Therefore, the relative effect of any variation in a hypothetical radio jet should be small compared with the variations in the H II region flux.

A further limitation of this study is that accretion onto the protostars is not well resolved, since the maximum cell resolution (98 AU) corresponds to a scale of the order of the inner accretion disk [161].

5.6   Conclusions

We performed an analysis of the radio-continuum variability in H II regions that appear in the radiation-hydrodynamic simulations of massive-star formation presented in Peters et al. [161]. The ultimate fate of ultracompact and hypercompact H II regions is to expand, but during their evolution they flicker due to the complex interplay of the inner ionized gas and the outer neutral gas. The radio-luminosities of the H II regions formed by the accreting protostars in our simulations are in agreement with those of observational radio surveys, except for the most luminous of the observed H II regions. We show that H II regions are highly variable in all timescales from 10 to 10^4 year, and estimate that at least 10% of observed ultracompact and hypercompact H II regions should exhibit flux variations larger than 10% for time intervals longer than about 10 year.

Chapter 6
Conclusions

6.1 General Conclusions

In this thesis we have performed radio to (sub)mm observations (Chaps. 2 and 3) in continuum, molecular lines, and hydrogen recombination lines that reveal the structure and dynamics of two MSFRs with very high luminosities ($L > 10^5 \, L_\odot$) that are in the process of forming stars at the upper end of the mass function. We have also analyzed archive observations in a third object (Chap. 4) that reveal that its radio-continuum emission is variable, and analyzed state-of-the-art numerical simulations of the formation of massive stars in clusters (Chap. 5) that explain several of the observed features.

Our main conclusions can be summarized as follows:

1. It appears that the most massive stars in the Galaxy ($M_\star > 15 \, M_\odot$) form by similar processes than low-mass stars, i.e., accretion of environment gas. However, ionization needs to be taken into account.
2. The onset of ionization does not appear to stop accretion initially. Infall and rotation in disk-like structures continues during part of the HC H II region stage. Analytical models and simulations suggest that accretion may continue all the way to the star in a partially-ionized accretion flow. However, conclusive observations are still not available.
3. Regardless of whether accretion continues to the star or not. Observations and models show that the youngest H II regions are not freely-expanding Strömgren spheres, but the inner, ionized part of a larger-scale accretion flow.
4. The molecular cores (scales of ~0.1 pc) that harbor disks, toroids, and HC H II regions do not appear to be isolated from their environment as considered in the so-called monolithic-collapse models of massive star formation. Rather, they appear to be coupled to their clump (pc-scale) environment. It is not possible to say without doubt that the clump-scale gas eventually reaches the star, but observations suggest that competitive accretion may play a significant role, at least in MSFRs that are forming very massive stars.

R.J. Galván-Madrid, *On the Formation of the Most Massive Stars in the Galaxy*,
Springer Theses, DOI 10.1007/978-1-4614-3308-8_6,
© Springer Science+Business Media New York 2012

Appendix A
Radio and (Sub)millimeter Interferometers

A.1 Aperture Synthesis

The angular resolution θ of a diffraction-limited telescope of diameter D at wavelength λ is $\theta \approx \lambda/D$. A 10-m class optical telescope can in principle achieve an angular resolution as good as $\theta \approx 0.01''$. However, in the radio regime (cm wavelengths) such a telescope would only have $\theta > 100''$. This problem can be circumvented with interferometry, in which each telescope can be considered as a small section in the aperture of a telescope of diameter B, the separation between elements. The technique of combining several telescopes into an interferometer and producing images from it is known as "aperture synthesis", or "synthesis imaging", and was developed first in the radio part of the spectrum because of the feasibility of combining signals electronically. Further discussion of this technique can be found in Wilson et al. [219] and Thompson et al. [199].

Consider a plane electromagnetic wave of amplitude E from a very distant source. This wave induces the voltages U_1 and U_2 in two antennae

$$U_1 \propto E \exp[i\omega t], \tag{A.1}$$

$$U_2 \propto E \exp[i\omega(t - \tau)], \tag{A.2}$$

where τ is the delay caused by extra path δl that the wave needs to travel to reach the second antenna. A schematic of a two-element interferometer is in Fig. A.1.

In a correlator the two signals are convolved (multiplied and integrated in time) producing an output:

$$R(\tau) \propto \int_0^T U_1 U_2 dt \propto E^2 \exp[i\omega\tau]. \tag{A.3}$$

For astronomical observations, τ varies with time due to Earth's rotation.

R.J. Galván-Madrid, *On the Formation of the Most Massive Stars in the Galaxy*, Springer Theses, DOI 10.1007/978-1-4614-3308-8, © Springer Science+Business Media New York 2012

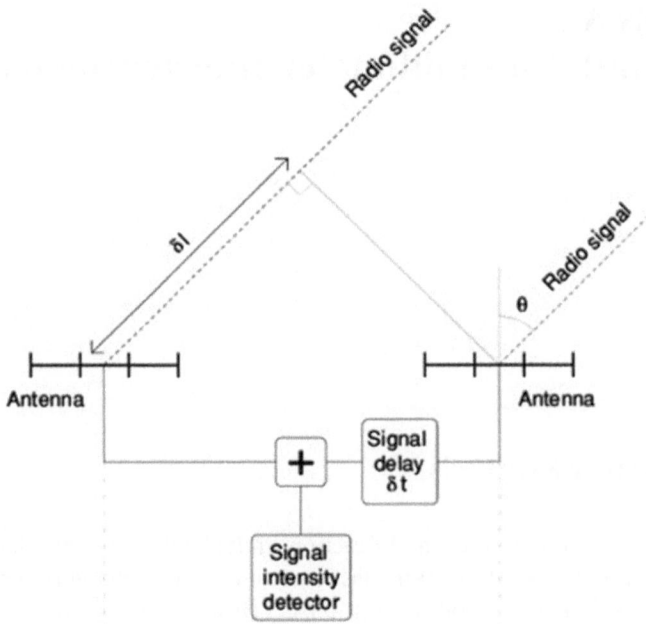

Fig. A.1 Schematic of a two-element radio interferometer

Now lets consider an object with a vector direction **s** from the antennae to its position on the sky, and a brightness distribution $I_V(\mathbf{s})$. The collecting area is $A(\mathbf{s})$, and **B** is the baseline vector joining the two antennae. The output from the correlator when the two-element interferometer stares at the source centered in **s** and covering the angle Ω is

$$R(\mathbf{B}) = \int_{\Omega} A(\mathbf{s}) I_V(\mathbf{s}) \exp\left[i2\pi v \left(\frac{1}{c}\mathbf{B}\cdot\mathbf{s} - \tau\right)\right] d\Omega \Delta v, \qquad (A.4)$$

$R(\mathbf{B})$ is known as the visibility function. In an array of many elements, every element pair with baseline $\mathbf{B}_{i,j}$ produces a visibility measurement $R_{i,j}$ per integration time.

A convenient coordinate system in the plane of the sky can be defined. Let $\sigma = (x, y)$ be a vector with origin in some reference within the source $\mathbf{s}_0$, then the normalized projection of **B** on the source plane is the (u, v) vector

$$(u, v) = \frac{\mathbf{B}\cdot\sigma}{\lambda}, \qquad (A.5)$$

and the complex visibility function (A.4) can be rewritten as:

$$V(u, v) = \int\int A(x, y) I(x, y) \exp[i2\pi(ux + vy)] dx dy \qquad (A.6)$$

Finally, the brightness distribution of the source on the sky is obtained from the Fourier transform of the visibility function $V(u,v)$ and dividing by the beam response $A(x,y)$:

$$I(x,y) = \frac{1}{A(x,y)} \int \int V(u,v) \exp[-i2\pi(ux+vy)]dudv \qquad (A.7)$$

A.2   Visibility Calibration

Visibilities need to be calibrated to remove instrumental and atmospheric effects. These factors usually depend on individual antennae or in antenna pairs. The relation of the uncalibrated visibility $V(u,v)_{\text{uncal}}$ to the true visibility $V(u,v)_{\text{true}}$ is :

$$V(u,v)_{\text{uncal}} = G_{i,j}(t)V(u,v)_{\text{true}}, \qquad (A.8)$$

where the complex gain factor $G_{i,j}(t)$ for each (u,v) measurement for each baseline defined by antennae i and j may depend on time t.

To calibrate $G_{i,j}(t)$, a calibrator source of known structure and flux is observed interspersed in time altogether with the target source. It is convenient to pick as calibrator a point-like, bright source, because the phase of an unresolved source at the phase center is zero. Unresolved quasars are typically a good choice to calibrate phases. If the selected quasar is not variable on the time-scale of a single observing run, it is also useful to calibrate the variation on the amplitude of $G_{i,j}(t)$ with time. However, the absolute amplitude scale still needs to be derived from either a source that is known to have a fixed flux on timescales of years to decades, or from a source for which a very reliable model of the emission exists. The flux scale in mm and submm observations is usually derived from models of the thermal emission of bodies in the Solar system. For centimeter observations, standard flux quasars are typically used, but their flux is ultimately derived from observations of Solar-system objects.

The visibility of a point-like calibrator source with flux S_s at the phase center is

$$V_c(u,v) = G_{i,j}(t)S_s \qquad (A.9)$$

Once the complex gain factors from (A.9) are known, they are applied using (A.8) to retrieve the true visibility of the target source.

The baseline-based gain factor can be decomposed in antenna gain factors as

$$G_{i,j}(t) = g_i(t)g_j^*(t) \qquad (A.10)$$

The more antennae an interferometer has, the more redundancy there is for determining the individual-antenna gains, facilitating calibration.

Appendix B
Molecular-Line Emission

B.1 Molecular Transitions

Molecules are more complex than atoms and their line emission can be produced by several different processes. The three main classes of molecular transitions are:

- Transitions between electronic levels. The corresponding spectral lines are typically in the visual and UV wavelength range.
- Vibrational transitions that arise from oscillations of the nuclei. These lines are typically in the IR.
- Rotational transitions that arise from rotational motions of the nuclei. Lines are mostly at sub(mm) and cm wavelengths. These are the transitions that are relevant to the instruments used in this thesis.

The energy levels E_{rot} of rotational transitions are defined by the rotational quantum number $J = 0, 1, 2, \ldots$:

$$E_{rot} = \frac{\hbar^2}{2\Theta} J(J+1),\tag{B.1}$$

where $\hbar$ is the reduced Planck constant and Θ is the moment of inertia of the molecule. Equation B.1 is only valid for rigid linear molecules. If a molecule can be centrifugally stretched the energy levels also depend on the stretching constant D as:

$$E_{rot} = \frac{\hbar^2}{2\Theta} J(J+1) - hD[J(J+1)]^2\tag{B.2}$$

Since dipole radiative transitions between rotational levels only occur in molecules with a permanent dipole moment, non-polar molecules like H_2, N_2, and O_2 are typically difficult to detect.

The structure of energy levels is more complicated ("hyperfine splitting") for molecules that have a non-zero electric dipole or magnetic quadrupole. A new

R.J. Galván-Madrid, *On the Formation of the Most Massive Stars in the Galaxy*, Springer Theses, DOI 10.1007/978-1-4614-3308-8,
© Springer Science+Business Media New York 2012

quantum number $F = J + I$ is conserved, where I is the nuclear spin. Examples are molecules containing a ^{14}N atom such as HCN, HNC, and HC$_3$N.

B.2   Carbon Monoxide

Carbon Monoxide (CO) is the most abundant linear molecule and it is widely used to trace molecular gas across the Universe. In this thesis we have observed the $J = 2 - 1$ transition at $v = 230.538$ GHz. The column density in the lower level of a linear molecule is given by [200, 219]:

$$N(J) = 93.5 \frac{g_J v^3}{g_{J+1} A_{J+1,J}} \frac{1}{1 - \exp[-4.80 \times 10^{-2} v / T_{ex}]} \int \tau dv, \qquad (B.3)$$

where the frequency v is in GHz, the linewidth dv in km s^{-1}, g_J and g_{J+1} are the statistical weights of the lower and upper levels respectively, $A_{J+1,J}$ is the spontaneous-emission Einstein coefficient, T_{ex} is the excitation temperature between the energy levels, and τ is the optical depth.

If the level populations are in LTE, the total CO column density is given by the Boltzmann equation:

$$N(\text{tot}) = N(J) \frac{Z}{2J+1} \exp\left[\frac{hB_e J(J+1)}{kT}\right], \qquad (B.4)$$

where B_e is the rotational constant and the partition function Z can be approximated by

$$Z \approx kT / hB_e, \text{ for } hB_e \ll kT \qquad (B.5)$$

Once the column density of CO is known, the column of molecular hydrogen can be estimated if the so-called $X(\text{CO})$ conversion factor is known. It is usually assumed that there are $\approx 10^4$ H$_2$ molecules for every CO molecule.

For massive star formation regions, CO is often optically thick at least in the central part of the spectrum. The same CO transition in an isotopologue like ^{13}CO or C^{18}O can be used to measure τ by means of the ratio of line intensities:

$$\frac{I(^{12}\text{CO})}{I(^{13}\text{CO})} = \frac{1 - \exp[-\tau_{J+1,J}]}{1 - \exp[-\tau_{J+1,J}]/\chi}, \qquad (B.6)$$

where the abundance ratio of ^{12}CO to ^{13}CO, χ, is of order 10^2 and known to vary with Galactocentric radius [218].

B.3 Ammonia Inversion Transitions

One of the molecules used through this thesis is ammonia (NH_3). This molecule is a "symmetric top", i.e., the molecule is symmetric in two of its three principal axes (see Fig. B.1). Taking z as the symmetry axis, and the angular momentum $J = (J_x, J_y, J_z)$, a symmetric-top molecule will have $\Theta_x = \Theta_y = \Theta_{pp}$ and $\Theta_z = \Theta_{pl}$. Its energy eigenvalues are:

$$E(J,K) = J(J+1)\frac{\hbar^2}{2\Theta_{pp}} + K^2\hbar^2\left(\frac{1}{2\Theta_{pl}} - \frac{1}{2\Theta_{pp}}\right),\tag{B.7}$$

where K is the eigenvalue of J_z, $J = 0, 1, 2, \ldots$, and $K = 0, \pm 1, \pm 2, \ldots \pm J$. The transitions that we use in this work are the hyperfine transitions where (J, K) remains constant after the N nucleus tunnels across the H ring. This so-called "inversion" transitions are at wavelengths of around 1.3 cm and are observable with the Very Large Array (VLA). Each (J, K) main line is accompanied by two pairs of hyperfine satellite components.

When at least one satellite line is detected in addition to the main line for a given NH_3 hyperfine transition (either in emission or absorption), the optical depth of the gas can be calculated. Assuming LTE, as well as a single filling factor and excitation temperature for either two main/satellite or two main(J, K)/main(J', K') pairs, the column density of ammonia N_{NH_3} and the rotational temperature of the gas T_{rot} can be estimated. Further details can be found in Ho and Townes [87] and Mangum and Wooten [141].

The optical depth of the main line $\tau(J, K, m)$ in a given channel is obtained by solving the equation:

$$\frac{T_B(J,K,m)}{T_B(J,K,s)} = \frac{1 - \exp[-\tau(J,K,m)]}{1 - \exp[-a\tau(J,K,m)]},\tag{B.8}$$

where a is the intrinsic relative intensity of the satellite (s) with respect to the main (m) line and T_B are the brightness temperatures.

Fig. B.1 3D structure of the ammonia NH_3 molecule. Nitrogen is in *blue* and Hydrogen in *white*. Credit Ben Mills

The column density $N(J,K)$ of ammonia molecules in a given (J,K) state is related to the velocity-integrated optical depth by:

$$N(J,K) = \frac{3hJ(J+1)}{8\pi^3\mu^2K^2}\left[\frac{1+\exp(-h\nu/kT_{ex})}{1-\exp(-h\nu/kT_{ex})}\right]\int\tau(J,K)d\nu \qquad (B.9)$$

Given that $h\nu \ll kT_{ex}$, and introducing $I_{J,K}$ as a factor to correct for the total optical depth of the main plus satellite lines $\tau(J,K) = I_{J,K}\tau(J,K,m)$, (B.9) can be re-expressed as

$$\left[\frac{N(J,K)}{cm^{-2}}\right] = 1.65 \times 10^{14}\frac{J(J+1)[T_{ex}/K]}{K^2[\nu/GHz]}I_{J,K}\tau_p(J,K,m)\left[\frac{\Delta V}{km\,s^{-1}}\right], \qquad (B.10)$$

where $\tau_p(J,K,m)$ is the peak optical depth of the main line for the (J,K) transition and ΔV is its FWHM, assuming a gaussian profile.

The total ammonia column density N_{NH_3} is then calculated by solving the equation:

$$\frac{N(J,K)}{N_{NH_3}} = \frac{(2J+1)S(J,K)}{Z}\exp\left[-\frac{E(J,K)}{T_{ex}(J',K';J,K)}\right], \qquad (B.11)$$

where the partition function Z is approximated by $Z \simeq 1.78 \times 10^6[T_{ex}^3/C_0B_0^2]^{1/2}$, with B_0 and C_0 being the rotational constants of ammonia, and $S(J,K)$ is the nuclear spin statistical weight.

By combining the Boltzmann equation with (B.10), the rotational temperature (i.e., the excitation temperature between K ladders) is obtained:

$$T_{rot}(J',K';J,K) = -\Delta E(J',K';J,K)$$

$$\left[\ln\left(\frac{(2J+1)J'(J'+1)K^2I_{J',K'}\tau_p(J',K',m)\Delta V(J',K')}{(2J'+1)J(J+1)K'^2I_{J,K}\tau_p(J,K,m)\Delta V(J,K)}\right)\right]^{-1}, \qquad (B.12)$$

T_{rot} gives a rough approximation to the kinetic temperature T_k of the molecular gas if collisions between the K ladders of ammonia can be ignored. However, to estimate T_k we apply the correction given by Danby et al. [54], where the collisional excitation of para- and ortho-NH$_3$ by para-H$_2$ $(J=0)$ was considered.

B.4   Methyl Cyanide Transitions

Methyl Cyanide (CH$_3$CN) is a symmetric-top molecule (see Fig. B.2) that is very useful to probe the physical conditions of dense molecular gas because it has groups of closely spaced lines with very different excitation. Given that for CH$_3$CN the

Fig. B.2 3D structure of the
methyl cyanide CH_3CN
molecule. Nitrogen is in *blue*,
Carbon in *black*, and
Hydrogen in *white*. Credit
Ben Mills

different K ladders are radiatively isolated (i.e., the relative populations of the K ladders are set by collisions), lines between them can provide us with a good estimate of the actual kinetic temperature of the gas at the observed scales. Since we have several lines and we are in the Rayleigh-Jeans regime, assuming LTE and low optical depth, we can obtain an estimate of the CH_3CN column density by the method outlined in the following paragraphs. For a more detailed discussion we refer the reader to Loren and Mundy [135] and Zhang et al. [232].

Making the assumptions mentioned above, the column density of CH_3CN molecules in the (J, K) (the upper J level of the emission line) state $N_{J,K}$ is related to the velocity-integrated line intensity by:

$$N_{J,K} = \left(\frac{3}{16\pi^3} \right) \left(\frac{\pi}{\ln 2} \right)^{1/2} \frac{(2J+1)}{S(J,K)} \left(\frac{k}{v\mu^2} \right) \int T_B dV, \qquad (B.13)$$

where $S(J,K)$ is the nuclear spin statistical weight and μ is the dipole moment of the molecule.

We have observed the $J = 12 - 11$ K-transitions of CH_3CN in the 1.3-mm band. In appropriate units (B.13) is rewritten as:

$$\left[\frac{N_{J,K}}{cm^{-2}} \right] = \frac{3.49 \times 10^{15}}{(v/GHz)(144 - K^2)} \int \frac{T_B dV}{K\ km\ s^{-1}} \qquad (B.14)$$

The relation between the population in a specific level and the total population is determined by the rotational temperature T_{rot} as given by:

$$\ln \left[\frac{N_{J,K}}{g_{J,K}} \right] = \ln \left[\frac{N_{CH_3CN}}{Z(T_{rot})} \right] - \frac{E_{J,K}}{k} \frac{1}{T_{rot}}, \qquad (B.15)$$

where Z is the partition function and $g_{J,K}$ is the statistical weight of the (J,K) level. The molecular parameters used are those of Boucher et al. [33]. Equation B.15 can be plotted in a "population diagram" for $K = 0, 1, 2, 3, 4, \ldots$ Then T_{rot} is determined from the slope of the fit, and N_{CH_3CN} is obtained from the intersection with the ordinate axis.

The procedure outlined above is only valid if lines are optically thin, which is often not the case at least for the low K numbers. One solution is to make the fit of the population diagram only using the higher K components [74]. A better solution is to drop off the assumption that lines are optically thin and fit simultaneously the multiple K lines solving for N_{CH_3CN} and T_{rot}, and taking into account the optical depth τ_v

$$\tau_v = \frac{c^2 N_{J,K} A_{J,K}}{8\pi v_0^2} \left(\exp[hv_0/kT_{rot}] - 1\right)\phi(v),\tag{B.16}$$

where v_0 is the rest frequency of the transition, and $\phi(v)$ is the line profile function, which we assume Gaussian. The Einstein spontaneous-emission coefficient of the upper level $A_{J,K}$ for rotational transitions of symmetric-top molecules is given by:

$$A_{J,K} = \frac{64\pi^4 v_0^3 \mu^2}{3hc^3} \frac{J^2 - K^2}{J(2J+1)}\tag{B.17}$$

Finally, in LTE, the level populations follow a Boltzmann distribution and the total column density N_{CH_3CN} is

$$\frac{N_{J,K}}{N_{CH_3CN}} = \frac{g_{J,K}}{Z(T_{rot})}\exp[-E_{J,K}/kT_{rot}]\tag{B.18}$$

Following Araya et al. [8], we use the partition function $Z(T_{rot})$

$$Z(T) = \frac{3.89[T/K]^{1.5}}{\left(1 - \exp[-524.8/[T/K]]\right)^2},\tag{B.19}$$

which includes contributions to the total population from the first vibrationaly-excited mode and full spin degeneracy factors [33].

Appendix C
Ionized-Gas Emission

C.1 Free–Free Continuum Emission

An ionized plasma emits continuum radiation due to accelerations caused by Coulomb interactions in an ensemble of charged particles. For typical interstellar conditions only the electrons suffer significant acceleration and radiate. A detailed derivation of the results used here can be found in Rybicki and Lightman [178]. The optical depth of this "free–free" radiation at frequency v is given by

$$\tau_v \sim 8.2 \times 10^{-2} \left[\frac{T_e}{K}\right]^{-1.35} \left[\frac{v}{GHz}\right]^{-2.1} \left[\frac{EM}{pc\ cm^{-6}}\right], \tag{C.1}$$

where T_e is the electron temperature and EM, the emission measure, is the integral of the electron density squared along the line of sight $EM = \int n_e^2 dl$. Note that the free–free emission gets optically thinner at higher frequencies.

Since for H II regions $T_e \sim 10^4$ K, the Rayleigh-Jeans limit $hv \ll kT$ is valid from radio to (sub)mm wavelengths. The intensity of an ionization-bounded (i.e., the ionization does not get to infinity) H II region along any given line of sight is

$$I_v = \frac{2v^2}{c^2} kT_B, \tag{C.2}$$

where $T_B = T_e(1 - \exp[-\tau_v])$ is the brightness temperature of the radiation.

Integrating over the solid angle Ω subtended by the H II region. The free–free continuum has the following spectral indices with frequency

$$S_v \propto v^{-0.1}, \text{ for } \tau_v \ll 1, \tag{C.3}$$

$$S_v \propto v^2, \text{ for } \tau_v \gg 1 \tag{C.4}$$

R.J. Galván-Madrid, *On the Formation of the Most Massive Stars in the Galaxy*, Springer Theses, DOI 10.1007/978-1-4614-3308-8, © Springer Science+Business Media New York 2012

Compact and most UC H II regions show this spectral break at frequencies ranging from 1 to 10 GHz. However, some UC H II regions and most HC H II regions with $n_e > 10^5 \, \mathrm{cm}^{-3}$ are partially optically thick due to density gradients and/or clumpiness (see discussions in Chaps. 2 and 4), causing intermediate spectral indices $S_v \sim 1$ all the way from 1 GHz to frequencies above 100 GHz.

C.2   Radio Recombination Lines

In a plasma of ionized hydrogen, electrons recombine with nuclei and get ionized back again continuously. While recombining, electrons cascade down from energy levels with higher principal quantum numbers n to lower levels. Transitions from $n + 1 \rightarrow n$ have the highest probability. From $n \sim 20$ to infinity they appear at sub(mm) to radio frequencies, producing the so-called radio recombination lines (RRLs).

The profiles of RRLs are given by three processes: thermal broadening due to the gas temperature T_e, "pressure" broadening due to distortion of the quantized energy levels caused by collisions, and dynamical broadening due to spatially unresolved motions in the form of either turbulence or ordered flows.

Pressure broadening hampers the understanding of the dynamics from RRLs, but the line-width due to this effect decreases rapidly with frequency as $\Delta v_{\mathrm{pr}} \propto v^{-4}$ [114]. Thermal broadening also makes dynamical studies difficult, but the thermal linewidth is fixed with frequency at $\Delta v_{\mathrm{th}} \approx 19 \, \mathrm{km \, s}^{-1}$ for $T_e = 8,000 \, \mathrm{K}$. Therefore, the best way to obtain the dynamics of the ionized gas is through a combination of continuum and line measurements, where at least one line observation is done at high frequencies (e.g., with the VLA at 7 mm, or with the SMA). This method is employed and further discussed in Chap. 2.

References

1. Abel T, Bryan GL, Norman ML (2002) Science 295:93
2. Acord JM, Churchwell E, Wood DOS (1998) ApJ Lett 495:107
3. Adams FC, Lada CJ, Shu FH (1987) ApJ 312:788
4. Afflerbach A, Churchwell E, Acord JM, Hofner P, Kurtz S, Depree CG (1996) ApJ Suppl 106:423
5. Anderson LD, Bania TM (2009) ApJ 690:706
6. Andrews SM, Wilner DJ, Hughes AM, Qi C, Dullemond CP (2009) ApJ 700:1502
7. Anglada G, Villuendas E, Estalella R, Beltrán MT, Rodríguez LF, Torrelles JM, Curiel S (1998) AJ 116:2953
8. Araya E, Hofner P, Kurtz S, Bronfman L, DeDeo S (2005) ApJ Suppl 157:279
9. Arce HG, Shepherd D, Gueth F, Lee C-F, Bachiller R, Rosen A, Beuther H (2007) In: Reipurth B, Jewitt D, Keil K (eds) Protostars and planets V. University of Arizona Press, Tucson, p 245
10. Arce HG, Santiago-García J, Jørgensen JK, Tafalla M, Bachiller R (2008) ApJ Lett 681:21
11. Arthur SJ, Hoare MG (2006) ApJ Suppl 165:283
12. Arthur SJ, Henney WJ, Mellema G, De Colle F, Vázquez-Semadeni E (2011) MNRAS 414:1747
13. Avalos M, Lizano S, Rodríguez LF, Franco-Hernández R, Moran JM (2006) ApJ 641:406
14. Avalos M, Lizano S, Franco-Hernández R, Rodríguez LF, Moran JM (2009) ApJ 690:1084
15. Ballesteros-Paredes J, Hartmann L, Vázquez-Semadeni E (1999) ApJ 527:285
16. Ballesteros-Paredes J, Klessen RS, Mac Low M-M, Vázquez-Semadeni E (2007) In: Reipurth B, Jewitt D, Keil K (eds) Protostars and planets V. University of Arizona Press, Tucson, p 63
17. Baobab Liu H, Ho PTP, Zhang Q, Keto E, Wu J, Li H (2010) ApJ 722:262
18. Bate MR, Bonnell IA (2005) MNRAS 356:1201
19. Becker RH, White RL, Helfand DJ, Zoonematkermani S (1994) ApJ Suppl 91:347
20. Beltrán MT, Cesaroni R, Neri R, Codella C, Furuya RS, Testi L, Olmi L (2004) ApJ Lett 601:187
21. Beltrán MT, Cesaroni R, Neri R, Codella C, Furuya RS, Testi L, Olmi L (2005) A&A 435:901
22. Beltrán MT, Cesaroni R, Codella C, Testi L, Furuya RS, Olmi L (2006) Nature 443:427
23. Beltrán MT, Cesaroni R, Moscadelli L, Codella C (2007) A&A 471:L13
24. Beltrán MT, Cesaroni R, Neri R, Codella C (2011) A&A 525:151
25. Bergin EA, Tafalla M (2007) ARA&A 45:339
26. Beuther H, Shepherd D (2005) In: Kumar MSN (ed) Cores to clusters: star formation with next generation telescopes. Springer, Dordrecht, p 105
27. Beuther H, Schilke P, Gueth F, McCaughrean M, Andersen M, Sridharan TK, Menten KM (2002) A&A 387:931

R.J. Galván-Madrid, *On the Formation of the Most Massive Stars in the Galaxy*,
Springer Theses, DOI 10.1007/978-1-4614-3308-8,
© Springer Science+Business Media New York 2012

28. Beuther H, Schilke P, Sridharan TK, Menten KM, Walmsley CM, Wyrowski F (2002) A&A 383:892
29. Beuther H, Schilke P, Stanke T (2003) A&A 408:601
30. Beuther H, Churchwell EB, McKee CF, Tan JC (2007) In: Reipurth B, Jewitt D, Keil K (eds) Protostars and planets V. University of Arizona Press, Tucson, p 165
31. Bonnell IA, Bate MR (2006) MNRAS 370:488
32. Bonnell IA, Bate MR, Vine SG (2003) MNRAS 343:413
33. Boucher D, Burie J, Bauer A, Dubrulle A, Demaison J (1980) J Phys Chem Ref Data 9:659
34. Bromm V, Larson RB (2004) ARA&A 42:79
35. Brooks KJ, Garay G, Voronkov M, Rodríguez LF (2007) ApJ 669:459
36. Bunn JC, Hoare MG, Drew JE (1995) MNRAS 272:346
37. Burbidge EM, Burbidge GR, Fowler WA, Hoyle F (1957) Rev Mod Phys 29:547
38. Carrasco-González C, Rodríguez LF, Anglada G, Martí J, Torrelles JM, Osorio M (2010) Science 330:1209
39. Carrasco-González C, Rodríguez LF, Torrelles JM, Anglada G, González-Martín O (2010) AJ 139:2433
40. Carrasco-González C, Galván-Madrid R, Anglada G, Osorio M, D'Alessio P, Hofner P, Rodríguez L F, Linz H, Araya E (2012) To appear in ApJ Letters (arXiv:1205.3302C)
41. Cesaroni R, Felli M, Testi L, Walmsley CM, Olmi L (1997) A&A 325:725
42. Cesaroni R, Hofner P, Walmsley CM, Churchwell E (1998) A&A 331:709
43. Cesaroni R, Felli M, Jenness T, Neri R, Olmi L, Robberto M, Testi L, Walmsley CM (1999) A&A 345:949
44. Cesaroni R, Neri R, Olmi L, Testi L, Walmsley CM, Hofner P (2005) A&A 434:1039
45. Cesaroni R, Galli D, Lodato G, Walmsley CM, Zhang Q (2007) In: Reipurth B, Jewitt D, Keil K (eds) Protostars and planets V. University of Arizona Press, Tucson, p 197
46. Churchwell E (2002) ARA&A 40:27
47. Codella C, Testi L, Cesaroni R (1997) A&A 325:282
48. Curiel S, Ho PTP, Patel NA, Torrelles JM, Rodríguez LF, Trinidad MA, Cantó J, Hernández L, Gómez JF, Garay G, Anglada G (2006) ApJ 638:878
49. Cyganowski CJ, Whitney BA, Holden E, Braden E, Brogan CL, Churchwell E, Indebetouw R, Watson DF, Babler BL, Benjamin R, Gómez M, Meade MR, Povich MS, Robitaille TP, Watson C (2008) AJ 136:2391
50. Dale JE, Bonnell IA, Clarke CJ, Bate MR (2005) MNRAS 358:291
51. Dale JE, Bonnell IA, Whitworth AP (2007a) MNRAS 375:1291
52. Dale JE, Clark PC, Bonnell IA (2007b) MNRAS 377:535
53. D'Alessio P, Hartmann L, Calvet N, Franco-Hernández R, Forrest WJ, Sargent B, Furlan E, Uchida K, Green JD, Watson DM, Chen CH, Kemper F, Sloan GC, Najita J (2005) ApJ 621:461
54. Danby G, Flower DR, Valiron P, Schilke P, Walmsley CM (1988) MNRAS 235:229
55. Davies B, Lumsden SL, Hoare MG, Oudmaijer RD, de Wit W-J (2010) MNRAS 402:1504
56. de Pree CG, Mehringer DM, Goss WM (1997) ApJ 482:307
57. De Pree CG, Wilner DJ, Mercer AJ, Davis LE, Goss WM, Kurtz S (2004) ApJ 600:286
58. de Wit WJ, Hoare MG, Oudmaijer RD, Mottram JC (2007) ApJ Lett 671:169
59. de Wit WJ, Hoare MG, Oudmaijer RD, Lumsden SL (2010) A&A 515:45
60. Downes D, Wilson TL, Bieging J, Wink J (1980) A&A Suppl 40:379
61. Dunham MM, Evans NJ, Terebey S, Dullemond CP, Young CH (2010) ApJ 710:470
62. Evans NJ and the Spitzer c2d Legacy Team (2009) ApJ Suppl 181:321
63. Federrath C, Banerjee R, Clark PC, Klessen RS (2010) ApJ 713:269
64. Field GB, Blackman EG, Keto ER (2008) MNRAS 385:181
65. Fish VL, Reid MJ, Wilner DJ, Churchwell E (2003) ApJ 587:701
66. Franco J, Kurtz S, Hofner P, Testi L, García-Segura G, Martos M (2000) ApJ Lett 542:143
67. Franco-Hernández R, Rodríguez LF (2004) ApJ Lett 604:105
68. Franco-Hernández R, Moran JM, Rodríguez LF, Garay G (2009) ApJ 701:974

69. Fryxell B, Olson K, Ricker P, Timmes FX, Zingale M, Lamb DQ, MacNeice P, Rosner R, Truran JW, Tufo H (2000) ApJ Suppl 131:273
70. Furuya RS, Cesaroni R, Codella C, Testi L, Bachiller R, Tafalla M (2002) A&A 390:L1
71. Galván-Madrid R, Avila R, Rodríguez LF (2004) Rev Mex AA 40:31
72. Galván-Madrid R, Vázquez-Semadeni E, Kim J, Ballesteros-Paredes J (2007) ApJ 670:480
73. Galván-Madrid R, Rodríguez LF, Ho PTP, Keto E (2008) ApJ Lett 674:33
74. Galván-Madrid R, Keto E, Zhang Q, Kurtz S, Rodríguez LF, Ho PTP (2009) ApJ 706:1036
75. Galván-Madrid R, Zhang Q, Keto ER, Ho PTP, Zapata LA, Rodríguez LF, Pineda JE, Vázquez-Semadeni E (2010) ApJ 725:17
76. Galván-Madrid R, Peters T, Keto ER, Mac Low M-M, Banerjee R, Klessen RS (2011) MNRAS 416:1033
77. Garay G, Lizano S (1999) PASP 111:1049
78. Garay G, Reid MJ, Moran JM (1985) ApJ 289:681
79. Gibb AG, Hoare MG (2007) MNRAS 380:246
80. Girart JM, Rao R, Marrone DP (2006) Science 313:812
81. Girart JM, Beltrán MT, Zhang Q, Rao R, Estalella R (2009) Science 324:1408
82. Goddi C, Humphreys EML, Greenhill LJ, Chandler CJ, Matthews LD (2011) ApJ 728:15
83. Gordon MA, Sorochenko RL (2002) Radio recombination lines. Their physics and astronomical applications. Astrophysics and space science library. Kluwer, Dordrecht/Boston
84. Gritschneder M, Naab T, Walch S, Burkert A, Heitsch F (2009) ApJ Lett 694:26
85. Heitsch F, Hartmann LW, Slyz AD, Devriendt JEG, Burkert A (2008) ApJ 674:316
86. Ho PTP, Haschick AD (1986) ApJ 304:501
87. Ho PTP, Townes CH (1983) ARA&A 21:239
88. Ho PTP, Vogel SN, Wright MCH, Haschick AD (1983) ApJ 265:295
89. Ho PTP, Moran JM, Lo KY (2004) ApJ Lett 616:1
90. Hoare MG, Kurtz SE, Lizano S, Keto E, Hofner P (2007) In: Reipurth B, Jewitt D, Keil K (eds) Protostars and planets V. University of Arizona Press, Tucson, p 181
91. Hofner P, Churchwell E (1996) A&A Suppl 120:283
92. Hofner P, Cesaroni R, Olmi L, Rodríguez LF, Martí J, Araya E (2007) A&A 465:197
93. Hollenbach D, Johnstone D, Lizano S, Shu F (1994) ApJ 428:654
94. Ignace R, Churchwell E (2004) ApJ 610:351
95. Jaffe DT, Martín-Pintado J (1999) ApJ 520:162
96. Jaffe DT, Stier MT, Fazio GG (1982) ApJ 252:601
97. Jappsen A-K, Klessen RS, Larson RB, Li Y, Mac Low M-M (2005) A&A 435:611
98. Jiménez-Serra I, Martín-Pintado J, Caselli P, Martín S, Rodríguez-Franco A, Chandler C, Winters JM (2009) ApJ Lett 703:157
99. Jiménez-Serra I, Caselli P, Tan JC, Hernández AK, Fontani F, Butler MJ, van Loo S (2010) MNRAS 406:187
100. Johnstone D, Hollenbach D, Bally J (1998) ApJ 499:758
101. Jorgensen JK, Bourke TL, Myers PC, Di Francesco J, van Dishoeck EF, Lee C-F, Ohashi N, Schoier FL, Takakuwa S, Wilner DJ, Zhang Q (2007) ApJ 659:479
102. Kahn FD (1974) A&A 37:149
103. Keto ER (1990) ApJ 355:190
104. Keto ER (2002) ApJ 580:980
105. Keto E (2003) ApJ 599:1196
106. Keto ER (2007) ApJ 666:976
107. Keto E, Klaassen P (2008) ApJ Lett 678:109
108. Keto E, Wood K (2006) ApJ 637:850
109. Keto ER, Zhang Q (2010) MNRAS 406:102
110. Keto ER, Ho PTP, Haschick AD (1987) ApJ 318:712
111. Keto ER, Ho PTP, Reid MJ (1987) ApJ Lett 323:117
112. Keto ER, Ho PTP, Haschick AD (1988) ApJ 324:920
113. Keto ER, Welch WJ, Reid MJ, Ho PTP (1995) ApJ 444:765
114. Keto E, Zhang Q, Kurtz S (2008) ApJ 672:423

115. Klaassen PD, Wilson CD (2007) ApJ 663:1092
116. Klaassen PD, Wilson CD (2008) ApJ 684:1273
117. Klaassen PD, Wilson CD, Keto ER, Zhang Q (2009) ApJ 703:1308
118. Kraemer KE, Jackson JM (1995) ApJ Lett 439:9
119. Kratter KM, Matzner CD (2006) MNRAS 373:1563
120. Kratter KM, Matzner CD, Krumholz MR, Klein RI (2010) ApJ 708:1585
121. Krumholz MR, Klein RI, McKee CF (2007) ApJ 656:959
122. Krumholz MR, Klein RI, McKee CF, Offner SSR, Cunningham AJ (2009) Science 323:754
123. Kumar MSN, Grave JMC (2007) A&A 472:155
124. Kurtz S (2005) In: Cesaroni R, Felli M, Churchwell E, Walmsley M (eds) Massive star birth:
     a crossroads of astrophysics IAU symposium 227. Cambridge University Press, Cambridge,
     p 111
125. Kurtz S, Churchwell E, Wood DOS (1994) ApJ Suppl 91:659
126. Kurtz S, Cesaroni R, Churchwell E, Hofner P, Walmsley CM (2000) In: Mannings V, Boss
     AP, Russell SS (eds) Protostars and planets IV. University of Arizona Press, Tucson, p 299
127. Lada CJ, Lada EA (2003) ARA&A 41:57
128. Lada CJ, Muench AA, Rathborne J, Alves JF, Lombardi M (2008) ApJ 672:410
129. Lai S-P, Crutcher RM, Girart JM, Rao R (2001) ApJ 561:864
130. Larson RB (1981) MNRAS 194:809
131. Leurini S, Beuther H, Schilke P, Wyrowski F, Zhang Q, Menten KM (2007) A&A 475:925
132. Liu HB, Quintana-Lacaci G, Wang K, Ho PTP, Li Z-Y, Zhang Q, Zhang Z (2011) The origin
     of OB clusters: from 10 pc to 0.1 pc. ArXiv e-prints
133. Liu H-B, Zhang Q, Ho PTP (2011) ApJ 729:100
134. López-Sepulcre A, Codella C, Cesaroni R, Marcelino N, Walmsley CM (2009) A&A 499:811
135. Loren RB, Mundy LG (1984) ApJ 286:232
136. Lovas FJ (2004) J Phys Chem Ref Data 33:177
137. Lugo J, Lizano S, Garay G (2004) ApJ 614:807
138. Mac Low M, Klessen R (2004) Rev Mod Phys 76:125
139. Mac Low M, van Buren D, Wood DOS, Churchwell E (1991) ApJ 369:395
140. Mangum JG, Wootten A (1994) ApJ Lett 428:33
141. Mangum JG, Wootten A, Mundy LG (1992) ApJ 388:467
142. Martí J, Rodríguez LF, Reipurth B (1998) ApJ 502:337
143. Martínez-García EE, González-Lópezlira RA, Bruzual-A G (2009) ApJ 694:512
144. McKee CF, Ostriker EC (2007) ARA&A 45:565
145. McKee CF, Tan JC (2003) ApJ 585:850
146. Mezger PG, Henderson AP (1967) ApJ 147:471
147. Molinari S, Pezzuto S, Cesaroni R, Brand J, Faustini F, Testi L (2008) A&A 481:345
148. Molinari S and the Herschel Hi-GAL Consortium (2010) A&A 518:L100
149. Moscadelli L, Goddi C, Cesaroni R, Beltrán MT, Furuya RS (2007) A&A 472:867
150. Motte F, Andre P, Neri R (1998) A&A 336:150
151. Müller HSP, Schlöder F, Stutzki J, Winnewisser G (2005) J Mol Struct 742:215
152. Muzerolle J, Hartmann L, Calvet N (1998) ApJ 116:2965
153. Newman WI, Wasserman I (1990) ApJ 354:411
154. Nummelin A, Bergman P, Hjalmarson ÅFriberg P, Irvine WM, Millar TJ, Ohishi M, Saito S
     (2000) ApJ Suppl 128:213
155. Olmi L, Cesaroni R, Hofner P, Kurtz S, Churchwell E, Walmsley CM (2003) A&A 407:225
156. Ossenkopf V, Henning T (1994) A&A 291:943
157. Osterbrock DE (ed) (1989) Astrophysics of gaseous nebulae and active galactic nuclei.
     University Science Books, Mill Valley
158. Patel NA, Curiel S, Sridharan TK, Zhang Q, Hunter TR, Ho PTP, Torrelles JM, Moran JM,
     Gómez JF, Anglada G (2005) Nature 437:109
159. Perley RA, Schwab FR, Bridle AH (eds) (1989) Volume 6 of ASP conference series.
     Astronomical Society of the Pacific, San Francisco
160. Peters T, Banerjee R, Klessen RS (2008) Phys Scr T132:014026

161. Peters T, Banerjee R, Klessen RS, Mac Low M-M, Galván-Madrid R, Keto ER (2010) ApJ 711:1017
162. Peters T, Mac Low M-M, Banerjee R, Klessen RS, Dullemond CP (2010) ApJ 719:831
163. Peters T, Klessen RS, Mac Low M-M, Banerjee R (2010) ApJ 725:134
164. Peters T, Banerjee R, Klessen RS, Mac Low M-M (2011) ApJ 729:72
165. Plume R, Jaffe DT, and Evans NJ II (1992) ApJ Suppl 78:505
166. Qiu K, Zhang Q (2009) ApJ Lett 702:66
167. Qiu K, Zhang Q, Wu J, Chen H-R (2009) ApJ 696:66
168. Remijan A, Sutton EC, Snyder LE, Friedel DN, Liu S-Y, Pei C-C (2004) ApJ 606:917
169. Rengarajan TN, Ho PTP (1996) ApJ 465:363
170. Roberts WW (1969) ApJ 158:123
171. Rodón JA, Beuther H, Megeath ST, van der Tak FFS (2008) A&A 490:213
172. Rodríguez LF (1997) In: Reipurth B, Bertout C (eds) Herbig-Haro flows and the birth of stars. IAU symposium 182. Kluwer, Dordrecht, p 83
173. Rodríguez LF, D'Alessio P, Wilner DJ, Ho PTP, Torrelles JM, Curiel S, Gómez Y, Lizano S, Pedlar A, Cantó J, Raga AC (1998) Nature 395:355
174. Rodríguez LF, Garay G, Brooks KJ, Mardones D (2005) ApJ 626:953
175. Rodríguez LF, Moran JM, Franco-Hernández R, Garay G, Brooks KJ, Mardones D (2008) AJ 135:2370
176. Rodríguez LF, Zapata LA, Ho PTP (2009) ApJ 692:162
177. Rosolowsky EW, Pineda JE, Foster JB, Borkin MA, Kauffmann J, Caselli P, Myers PC, Goodman AA (2008) ApJ Suppl 175:509
178. Rybicki GB, Lightman AP (1979) Radiative processes in astrophysics. Wiley, New York
179. Salpeter EE (1955) ApJ 121:161
180. Sewilo M, Churchwell E, Kurtz S, Goss WM, Hofner P (2004) ApJ 605:285
181. Sewilo M, Churchwell E, Kurtz S, Goss WM, Hofner P (2008) ApJ 681:350
182. Shepherd DS, Claussen MJ, Kurtz SE (2001) Science 292:1513
183. Shu FH, Milione V, Gebel W, Yuan C, Goldsmith DW, Roberts WW (1972) ApJ 173:557
184. Shu FH, Adams FC, Lizano S (1987) ARA&A 25:23
185. Shu F, Najita J, Ostriker E, Wilkin F, Ruden S, Lizano S (1994) ApJ 429:781
186. Smith RJ, Longmore S, Bonnell I (2009) MNRAS 400:1775
187. Sollins PK (2005) Accretion and outflow in massive star formation: observational studies at high angular resolution. PhD thesis, Harvard University, Cambridge
188. Sollins PK, Ho PTP (2005) ApJ 630:987
189. Sollins PK, Zhang Q, Keto E, Ho PTP (2005) ApJ Lett 624:49
190. Sollins PK, Zhang Q, Keto E, Ho PTP (2005) ApJ 631:399
191. Solomon PM, Rivolo AR, Barrett J, Yahil A (1987) ApJ 319:730
192. Spergel DN, Verde L, Peiris HV, Komatsu E, Nolta MR, Bennett CL, Halpern M, Hinshaw G, Jarosik N, Kogut A, Limon M, Meyer SS, Page L, Tucker GS, Weiland JL, Wollack E, Wright EL (2003) ApJ Suppl 148:175
193. Spitzer L (1978) Physical processes in the interstellar medium. Wiley, New York
194. Springel V, Yoshida N, (2001) White SDM. New Astron 6:79
195. Sridharan TK, Beuther H, Schilke P, Menten KM, Wyrowski F (2002) ApJ 566:931
196. Stier MT, Jaffe DT, Rengarajan TN, Fazio GG, Maxson CW, McBreen B, Loughran L, Serio S, Sciortino S (1984) ApJ 283:573
197. Tang Y-W, Ho PTP, Koch PM, Girart JM, Lai S-P, Rao R (2009) ApJ 700:251
198. Tenorio-Tagle G (1979) A&A 71:59
199. Thompson AR, Moran JM, Swenson GW Jr (eds) (2001) Interferometry and synthesis in radio astronomy, 2nd edn. Wiley, New York
200. Tielens AGGM (ed) (2005) The physics and chemistry of the interstellar medium. Cambridge University Press, Cambridge
201. Tobin JJ, Hartmann L, Looney LW, Chiang H-F (2010) ApJ 712:1010
202. Turner BE (1979) A&A Suppl 37:1
203. Vacca WD, Garmany CD, Shull JM (1996) ApJ 460:914

204. van der Tak FFS, Menten KM (2005) A&A 437:947
205. van der Tak FFS, van Dishoeck EF, Evans NJ II, Blake GA (2000) ApJ 537:283
206. van der Tak FFS, Tuthill PG, Danchi WC (2005) A&A 431:993
207. Vázquez-Semadeni E, Kim J, Shadmehri M, Ballesteros-Paredes J (2005) ApJ 618:344
208. Vázquez-Semadeni E, Gómez GC, Jappsen AK, Ballesteros-Paredes J, González RF, Klessen RS (2007) ApJ 657:870
209. Vázquez-Semadeni E, Gómez GC, Jappsen A-K, Ballesteros-Paredes J, Klessen RS (2009) ApJ 707:1023
210. Walsh AJ, Burton MG, Hyland AR, Robinson G (1998) MNRAS 301:640
211. Walsh AJ, Longmore SN, Thorwirth S, Urquhart JS, Purcell CR (2007) MNRAS 382:L35
212. Wang P, Li Z-Y, Abel T, Nakamura F (2010) ApJ 709:27
213. Ward-Thompson D, André P, Crutcher R, Johnstone D, Onishi T, Wilson C (2007) In: Reipurth B, Jewitt D, Keil K (eds) Protostars and planets V. University of Arizona Press, Tucson, p 33
214. Westerhout G (1958) Bull Astron Inst Neth 14:215
215. Wheeler JC, Sneden C, Truran JW Jr (1989) ARA&A 27:279
216. Williams JP, Mann RK, Beaumont CN, Swift JJ, Adams JD, Hora J, Kassis M, Lada EA, Román-Zúñiga CG (2009) ApJ 699:1300
217. Wilner DJ, Wright MCH, Plambeck RL (1994) ApJ 422:642
218. Wilson TL, Rood R (1994) ARA&A 32:191
219. Wilson TL, Rohlfs K, Hüttemeister S (eds) (2009) Tools of radio astronomy. Springer, Berlin
220. Wolfire MG, Cassinelli JP (1987) ApJ 319:850
221. Wood DOS, Churchwell E (1989) ApJ Suppl 69:831
222. Wu J, Evans NJ II (2003) ApJ Lett 592:79
223. Wu Y, Wei Y, Zhao M, Shi Y, Yu W, Qin S, Huang M (2004) A&A 426:503
224. Yorke HW, Sonnhalter C (2002) ApJ 569:846
225. Yorke HW, Welz A (1996) A&A 315:555
226. Young LM, Keto E, Ho PTP (1998) ApJ 507:270
227. Zapata LA, Palau A, Ho PTP, Schilke P, Garrod RT, Rodríguez LF, Menten K (2008) A&A 479:L25
228. Zapata LA, Ho PTP, Schilke P, Rodríguez LF, Menten K, Palau A, Garrod RT (2009) ApJ 698:1422
229. Zhang Q, Ho PTP (1995) ApJ Lett 450:63
230. Zhang Q, Ho PTP (1997) ApJ 488:241
231. Zhang Q, Wang K (2011) ApJ 733:26
232. Zhang Q, Ho PTP, Ohashi N (1998) ApJ 494:636
233. Zhang Q, Hunter TR, Sridharan TK (1998) ApJ Lett 505:151
234. Zhang Q, Hunter TR, Sridharan TK, Ho PTP (2002) ApJ 566:982
235. Zhang Q, Hunter TR, Beuther H, Sridharan TK, Liu S-Y, Su Y-N, Chen H-R, Chen Y (2007) ApJ 658:1152
236. Zhang Q, Wang Y, Pillai T, Rathborne J (2009) ApJ 696:268
237. Zhu Z, Hartmann L, Gammie C (2009) ApJ 694:1045
238. Zinnecker H, Yorke HW (2007) ARA&A 45:481